Growing Marijuana for Beginners and Use the Hydroponic System

How to grow Marijuana
by improving quantity and
quality even in small spaces

Garrick S. Thatcher

© Copyright 2020 by Garrick S. Thatcher.

All right reserved.

TABLE OF CONTENTS

Growing Marijuana for Beginners

Introduction	17
What Is Marijuana?	17
A History of Marijuana	18
The Legality of Marijuana Today	21
Chapter 1: Cannabis Basics	25
Cannabis Versus Marijuana Versus Hemp	25
CBD vs. THC	26
Tetrahydrocannabinol or THC	27
The Benefits of THC	28
Cannabidiol or CBD	35
The Benefits of CBD	36
The Mechanisms for Using Marijuana	46

Inhaling Through Smoking — 47

Inhaling Through Vaporizing — 50

Consuming Through Edibles — 50

Chapter 2: Understanding Cannabis Plants — 53

Indica, Sativa and Hybrid — 53

Common Strains and Their Characteristics — 56

    Indica 1: Granddaddy Purple — 57

    Sativa 1: Sour Diesel — 58

    Hybrid 1: Blue Dream — 58

    Hybrid 2: Strawberry Banana — 59

Identifying a Cannabis Plant — 60

Cannabis Plant Sex — 61

Chapter 3: Growing Marijuana — 65

Options for Growing Marijuana — 66

Growing Marijuana Inside Versus Outside — 66

Growing Marijuana Using Soil — 69

Growing Marijuana Using Hydroponics 69

Soil-Based Gardening Versus Hydroponics 70

Chapter 4: The Cannabis 73
Plant Life Cycle 73

Stage 1: The Seed Stage 74

When to Plant Your Seeds 78

Stage 2: Seed Germination 78

Stage 3: Seedling Growth 83

Stage 4: The Vegetative Stage 85

Vegetative: Training and Topping 85

Stage 5: The Pre-Flowering Stage 91

Stage 6: The Flowering Stage 92

Cannabis Plant Sex 92

Male Versus Female Cannabis Plants 93

The Flowering Stage Continued 93

Testing for Maturity 95

Drying the Buds	97

Chapter 5: The Logistics of Growing Marijuana 101

What You Need in Order to Get Started Growing Marijuana in Soil	101
What You Need to Get Started Using Hydroponics	104
The Cost of Growing Marijuana With Hydroponics	109
Hydroponics Equipment List	110
Additional Equipment	111

Chapter 6: Marijuana   113
Plant Maintenance   113

Maintenance That Needs to Be Done	113
Temperature	114
Oxygen	117
Water Levels	117

Lighting Cycles 119

Tools and Cultivation 119

Chapter 7: How to Avoid Pests and Disease 121

Common Marijuana Plant Pests 121

Common Marijuana Plant Diseases 123

How to Avoid and Prevent Pests 125

How to Deal With Disease 129

Chapter 8: How to Grow Healthy Cannabis Plants 133

Common Mistakes to Avoid When Growing Marijuana 133

Forgetting pH 133

Lighting Errors 134

Overfeeding Your Plants 135

Improper Nutrients 135

Tips and Tricks for First Time Hydroponics Growers   135

Chapter 9: What to Do Once Your Plants Have Grown   139

Cloning   139

Conclusion   143

Hydroponics

Introduction   149

Chapter 1: Welcome   151
to Hydroponics   151

What Is Hydroponics?   151

How Hydroponics Works   153

Chapter 2: Water or Soil?   157

Advantages of Hyrdoponic Gardening   158

    Disadvantages of Hydroponic Gardening   161

    Hydroponics: Yes or No   164

Chapter 3: Types 165
of Hydroponic Systems 165

    Drip System   166

    Flood and Drain System (EBB and Flow)   169

    Wick System   172

    Aeroponic System   177

    Deep Water Culture System   180

    Nutrient Film Technique System   183

Chapter 4: Materials 187
for Building a 187
Hydroponic System 187

    Grow Lights   188

    Submersible Pump   189

    Reservoir   190

Growing Tray ... 192

System for Nutrient Delivery ... 193

Air Pump ... 194

Growing Mediums ... 195

Chapter 5: Nutrients, Pest Control, and PH Balancing ... **203**

Nutrients ... 203

Pest Control ... 207

PH Balancing ... 215

Test Strips ... 218

Liquid Test Kits ... 218

Electronic Meter ... 219

Chapter 6: Ideal Plants for the Beginning Gardener ... **221**

Hydroponic Herbs ... 222

Hydroponic Flowers 226

Hydroponic Vegetables 230

Chapter 7: Ways to Succeed and Mistakes to Avoid 235

Ways to Succeed With Hydroponic Gardening 236

Mistakes to Avoid With Hydroponic Gardening 243

Common HydroPonic Problems and Their Solutions 246

Chapter 8: The Business of Hydroponics 253

Conclusion 259

Growing Marijuana for Beginners

The Complete Step-By-Step Guide for Beginners on Indoor and Outdoor Marijuana Cultivation for Big Buds

Garrick S. Thatcher

© **Copyright 2020 by Garrick S. Thatcher.**
All right reserved.

The work contained herein has been produced with the intent to provide relevant knowledge and information on the topic on the topic described in the title for entertainment purposes only. While the author has gone to every extent to furnish up to date and true information, no claims can be made as to its accuracy or validity as the author has made no claims to be an expert on this topic. Notwithstanding, the reader is asked to do their own research and consult any subject matter experts they deem necessary to ensure the quality and accuracy of the material presented herein.

This statement is legally binding as deemed by the Committee of Publishers Association and the American Bar Association for the territory of the United States. Other jurisdictions may apply their own legal statutes. Any reproduction, transmission or copying of this material contained in this work without the express written consent of the copyright holder shall be deemed as a copyright violation as per the current legislation in force on the date of publishing and subsequent time thereafter. All additional works derived from this material may be claimed by the holder of this copyright.

The data, depictions, events, descriptions and all other information forthwith are considered to be true, fair and accurate unless the work is expressly described as a work of fiction.

Regardless of the nature of this work, the Publisher is exempt from any responsibility of actions taken by the reader in conjunction with this work. The Publisher acknowledges that the reader acts of their own accord and releases the author and Publisher of any responsibility for the observance of tips, advice, counsel, strategies and techniques that may be offered in this volume.

INTRODUCTION

Marijuana is something that everybody knows about, but that few understand in-depth. This book will not only ensure that you understand the basics of marijuana and what it can be used for, but also that you will understand how to grow your own marijuana plants! In this introduction, we are going to ensure that we are all on the same page in terms of our basic cannabis and marijuana knowledge before diving into these topics throughout the rest of the book. After you finish reading the introduction, you will be ready to take on the first chapter with confidence.

What Is Marijuana?

Marijuana is the main topic that we will be discussing throughout this book, and we will explore it in-depth over the next several chapters. Marijuana is a term used to describe a type of plant. This plant species has gained immense popularity as well as media, political, and global attention over the past century- especially in the past fifty years. This plant contains medicinal benefits as well as intoxicating ones, which is part of the reason that it has become so popular in recent years.

Marijuana is the name given to the dried leaves, seeds, stems, and flowers of the cannabis plant. This plant can come in many different forms, and these forms can be used for many different purposes. The reason why marijuana is so popular amongst young people and also the reason why it is classified as a banned substance in much of the world is that the consumption of these dried flowers either by inhalation, ingestion, topical application or any other method leads to the feeling of being "high," classified as the pleasant feeling of relaxation, euphoria, and fatigue.

The cannabis plant can withstand a wide range of temperature conditions, water levels, and other weather-related variables. It is a very sturdy and resilient plant, which is likely why one of its nicknames is "weed". Weeds are notoriously pesky plants that keep coming back once you remove them from your garden, and seem to be able to withstand virtually any weather conditions.

A History of Marijuana

Knowing where marijuana came from and how it was discovered can help us to understand how it developed into the multi-million-dollar industry that it is today. We will begin by looking at where it was discovered and where it came from before delving deeper into its history.

Cannabis can be found growing in the wild in a variety of regions across the world. It is difficult to trace marijuana back to a specific location, time, or use, but scientists are hard at work trying to determine the exact origins of this wonderful plant. As a result of recent scientific investigation of fossil pollen, scientists suspect that marijuana is derived from the Tibetan Plateau. The fossils that led scientists to draw this conclusion demonstrate that the origins of the marijuana plant date back to this location in northwestern China, somewhere around 19.6 million years ago. They then began to determine that this plant traveled to Europe 6 million years ago and then to the eastern side of China 1.2 million years ago. It has also been found growing in the northern regions of Pakistan. There, it has been known to reach heights comparable to the first story of a high-rise building! These wild cannabis plants do not contain very high levels of the chemical that is responsible for the psychoactive effects of consuming cannabis, which is why it was not used as a recreational drug when it was first discovered. In the beginning of is use by humans, marijuana (or cannabis) was used as an herbal medicine instead of as a recreational drug and a means of getting high. These first uses of marijuana date back to somewhere around 500 BC, and this was reported to be somewhere in the central region of the continent of Asia.

Back in that time, the marijuana plant was harvested in the wild and grown in order to be used by humans as a means of pain relief herbal medicine as well as for other herbal medicine uses.

Over time, marijuana was found to have many more uses than simply medicinal pain relief, and this plant continued to be grown intentionally for some time. Eventually, it was then introduced into the ecosystems of Africa and Europe so that these regions could benefit from this plant as well.

These original strains of marijuana plants were only mildly psychoactive, and this could be the reason why it was used primarily for purposes other than as a recreational drug. There are some researchers who believe that the ancient peoples who used these plants for their medicinal purposes knew about the psychoactive potential of marijuana, but there is more evidence supporting the fact that this was not its primary use. There are also some researchers who believe that there were some ancient peoples and cultures that developed and grew different strains of hemp plants intentionally, leading them to eventually develop plants containing higher levels of the psychoactive components within marijuana, called THC.

While the first uses of marijuana recreationally are much more recent than its first discovery, the plant was surely growing and blowing in the wind well before humans existed, let alone discovered it and decided to begin using it for their own medicinal and recreational purposes.

The Legality of Marijuana Today

The first reported uses of marijuana in America was somewhere around the time of its colonization, somewhere around 1492. This was reportedly when it began to be grown and cultivated in this part of the world. In colonial times in America, farmers were responsible for growing and harvesting marijuana plants. Since it is a plant that grows quickly and is very versatile and resilient, this plant was able to be grown and harvested in very large quantities. Eventually, this became the only way that marijuana was allowed to be grown, at least in the colonies of Connecticut, Massachusetts, and Virginia around the 1600s.

The difference in this part of the world when compared to Asia, is that the marijuana plant was grown to be used for fabric and textiles, as well as to make rope, and not for medicinal purposes.

In the Americas, people grew the hemp plant and made it useful for clothing as well as paper and even for food, as people began to understand that the seeds of the hemp plant were edible and actually a great source of protein, vitamins, and minerals. After its introduction in the Americas, marijuana gained a lot of attention and became controversial for racial and political reasons. As marijuana was introduced and then later controlled, there were very significant issues of race within the United States. This eventually led to the criminalization of marijuana in the USA, and this illegality remains today in the vast majority of states within the USA.

In 1937, there was something called the Marijuana Tax Act of 1937, which is a law created by the federal government that was put into place in order to criminalize marijuana in the United States. Further, this law placed a tax on the sale and the possession of any sort of hemp product. This limited the production and distribution of marijuana to industrial companies and uses only and began the punishment of anyone found with marijuana for any other uses. One day after this law was put into place, a farmer was charged with the sale of marijuana and was sentenced to 4 years of punishment for this action. His punishment included hard labor of a variety of sorts.

The War On Drugs began in 1970, when Richard Nixon, the President of the United States at the time, decided that the Marijuana Tax Act would be abolished and that marijuana would from then on be considered a Schedule 1 drug. Other drugs included in the Schedule 1 category included heroin, ecstasy, and LSD. Under this new Controlled Substances Act of 1970, Richard Nixon deemed marijuana a drug that lacked any medical uses, and that carried a high potential for drug abuse. It was considered from them on to be a "gateway drug," meaning that it was considered to be a drug that would lead people to other, more severe drug addictions or drug abuse such as heroin or cocaine.

Marijuana is a plant that comes with a deep history, and it is still, to this day, a heavily debated topic of discussion. This plant has endless possibilities and uses, and it is widely used and debated across the world because of this. To this day, there are many debates about whether or not it should be legal to grow, sell, and consume it due to its numerous benefits outside of getting people high. There is still a long way to go when it comes to this plant, but if you are lucky enough to live in a location where growing marijuana is legal, the rest of this book will teach you exactly how you can benefit from the many uses of this seemingly magical plant.

CHAPTER 1: CANNABIS BASICS

In this chapter, I am going to begin the book by sharing everything you need to know about Marijuana and its possible uses. After reading through the entirety of this chapter, you will be equipped with a foundation of knowledge about marijuana upon which we can build throughout the rest of the chapters in this book.

Cannabis Versus Marijuana Versus Hemp

The terms *cannabis, marijuana* and *hemp* are often used interchangeably, but it is necessary to understand the difference as we move forward in this book. By understanding this distinction, you will be able to fully grasp the topics in each chapter of this book. For this reason, in this section, I will outline for you the difference between Cannabis, Hemp, and Marijuana.

Hemp and Marijuana are two different forms of the Cannabis plant. Marijuana is what most people are speaking of when they use casual or "street" terms such as *weed, pot,* or *Mary Jane*. Marijuana is the type of cannabis that includes significant amounts of the compound THC, which we are going to learn about in the next section.

Hemp, however, is a different form of cannabis. Hemp is different from marijuana in its THC content. Hemp is cannabis that contains less than 0.3% THC. This means that it is still cannabis, but that it does not have significant levels of psychoactivity when consumed. This is because it is THC that is responsible for the effects known as a *high*. For this reason, hemp is often used for things such as clothing and for medicinal purposes, among other things.

Throughout this book, I will use the word *cannabis* when speaking about the plant in a general sense, and I will use the word *marijuana* when speaking about the form of the plant that is used as a recreational drug.

CBD vs. THC

In this subchapter, we are going to look a little deeper into the science of marijuana and the science behind the uses of marijuana. By learning more about the science behind this plant, you will be able to understand how it can benefit you and the mechanisms by which this happens in your body. This will also help you to understand why this plant is loved and cherished by so many.

There are two main components of marijuana that are responsible for the effects which make it such a desirable drug and therapy. These chemicals are called THC, which stands for *tetrahydrocannabinol* and CBD, which stands for *cannabidiol*. These chemicals act in different ways, and in this section, we will look at what makes them the chemical of choice when it comes to choosing a recreational drug.

Tetrahydrocannabinol or THC

THC, as you know by now, stands for tetrahydrocannabinol. There are many ways to consume THC; the most common way is through the marijuana flower. When THC is consumed by any method, it makes its way into the bloodstream. We will look at how this happens later on in this chapter. When THC reaches the bloodstream and the brain, it produces the effects that are known as the "high," those feelings of euphoria and relaxation. There are receptors in the brain that THC can bind to, and this produces the high.

THC comes with some side-effects in high amounts, though these side-effects are temporary, meaning that they are mostly present only for the duration of the high. These side-effects can include an increase in heart rate, lowered coordination, memory impairment, dry mouth, red eyes, and a reduction in reaction time.

The side-effects that THC produces are a direct result of its psychoactive effects on the brain. We will look at the mechanisms by which this works in the next section of this chapter. THC is not known to lead to death for any reason, but it can lead to long-term psychiatric changes or issues as a result of long-term use by adolescents whose brains are not finished growing and developing.

The Benefits of THC

While many people consider the high feeling that THC gives them to be a positive or a benefit, many others do not see it this way. In this section, we will look at the other benefits that THC can provide you with, aside from a feeling of being high.

THC is known to help with the following, in addition to the feeling of being "high;"

- Insomnia

Marijuana has been shown to lead to fatigue, which makes it a good treatment for insomnia. Many people turn to marijuana to help them fall asleep each night. Further, because of the relaxation and euphoria that marijuana can lead to, this can help people suffering from insomnia to relax and fall asleep with more ease.

- Stress

The euphoria associated with THC consumption can lead to a reduction in stress as it helps people to see things in a more positive light than they may have before consuming THC. Further, the relaxation and fatigue that people experience when they consume cannabis is very helpful in reducing stress levels.

- Pain

THC is a proven remedy for all sorts of different pain that people may experience. THC was first legalized solely for medicinal purposes, one of the main purposes being pain reduction. For people suffering from chronic illnesses, consuming THC is a proven, natural method to relieve their pain. There are different strains of marijuana, each containing different levels of THC. These different strains are beneficial for different reasons and different types of pain, so people wishing to use THC for pain relief can begin to experiment with different strains depending on what they are looking for. There are three different types of pain that a person may experience. The first is called *nociceptive pain*. This type of pain is the type that a person would experience after some type of damage to the body, such as breaking an arm or having arthritis (since this results from inflammation and damage to the joints).

The second type of pain is called *neuropathic pain,* and this type of pain is the type a person would experience as a result of nerve pain or nerve-related issues in their body. An example of this could be carpal tunnel syndrome or any type of nerve damage. The third and final type of pain is called *central pain.* Central pain is a type of pain that stems from the central nervous system, which includes the brain, the brainstem, and the spinal cord. This pain results from some type of damage to one of these areas or to a type of dysfunction of one of these areas. The problems that cause this pain can range from a stroke to brain trauma to a tumor to Parkinson's disease and to fibromyalgia. This type of pain is due to more severe issues since it involves the brain and the spinal cord. The type of pain that you experience can determine the type of remedy you need.

For nociceptive pain (injury pain), the pain is caused by damage to an area of the body and the inflammation that occurs in that area as a response to the damage. In order to relieve this type of pain, it is necessary to reduce inflammation. THC aids in reducing inflammation, which, in turn, reduces the pain associated with the injury. Further, the pain signals travel from the site of injury to the brain, where your brain interprets it as pain. THC also acts to slow or lessen the traveling of these pain signals in order to reduce a person's perception of their pain.

Neuropathic pain or pain associated with the body's nerves is the type of pain that arises from damage to the nervous system. This type of pain is most often caused by the pinching of nerves, which can cause damage eventually, the injury to a nerve, or by some type of stabbing that causes nerve damage. The difference between this type of pain and nociceptive pain is that this type does not involve inflammation, so it is difficult to treat with common drugs that usually target inflammatory cells. The reason that THC is such a beneficial treatment for this type of pain is that it leads to the reduction of pain in different ways than most pharmaceutical products do. THC leads to a reduction in pain related to nerve damage by protecting the nerve cells themselves, as well as increasing the person's mood and feelings of positivity. These two effects, when combined, are powerful neuropathic pain relievers.

Central pain is a relatively new term as it is a relatively new discovery in the science world. Central pain is tricky as it is present in the absence of any visible injury or any real source. Fibromyalgia is an example of this type of pain. Fibromyalgia is a sort of general pain that can be experienced all over the body. There is no root cause of fibromyalgia, and this is why it is an interesting field of study as of late. This disease is thought to be caused by dysfunction in the way that the brain processes pain signals.

It is very difficult to find treatments for this type of pain, but THC has proven to be extremely beneficial in treating it. This could be due to the way that it slows and dulls the pain signals as they travel to the brain or due to the change in mood that accompanies THC consumption. The mechanisms of action on central pain, in particular, are still largely unknown, but at this stage, it has shown to be a great option for treating this type of pain.

For example, a recent study conducted on people who were suffering from Multiple Sclerosis (MS) showed that treatment with THC led to a reduction in the pain that they regularly experienced due to muscle and joint stiffness and tightness, as well as the nerve pain that they experience. People suffering from MS experience regular nerve pain as their nerve cells lose their protective coating. This lack of protection leaves the nerve cells exposed and causes the person pain at the nerve-level.

- Nausea

It is not completely understood why yet, but THC consumption has been shown to reduce nausea. This can be beneficial for a variety of reasons, such as those who are going through chemotherapy or who suffer from nausea for any other reasons.

- Muscle Spasticity

Muscle Spasticity is the presence of involuntary muscle contractions that lead to spasms. This can happen for a variety of reasons that are not too serious, but for some people, it can become a recurring problem. In these individuals, much of the time, the treatment includes some combination of anti-inflammatory drugs, anti-anxiety medications, muscle relaxants, and other prescribed drugs such as these. The great thing about marijuana is that it has been shown to produce all of these effects in one. Instead of turning to a cocktail of prescribes pharmaceuticals, one can achieve all of these effects from a single treatment through the consumption of marijuana.

- Glaucoma

Glaucoma is a medical condition that involves the main nerve in the eye to be damaged for some reason or another. This is usually due to genetics, and it can progress to a point where a person experiences vision loss or even blindness. This happens because of an increase in the pressure within the eye, which eventually damages the *optic nerve.* The optic nerve, the main nerve within the eye, is responsible for sending visual messages from the eye to the brain to be interpreted. IF these messages cannot be sent to the brain to be interpreted, a person will lose the ability to see as the visual messages cannot reach their brain.

Marijuana consumption in patients with glaucoma reduces pressure within the eye, which can prevent any further damage to the optic nerve, prevent damage that has already occurred from getting worse, and can reduce the risk of developing lasting damage due to high-pressure levels within the eye.

One of the reasons why this is still a new area of study within the medical research community is the side-effects that go along with THC consumption must be considered in these patients, as they would need to be consuming THC every four hours or so since the pressure reduction in the eye caused by THC only lasts for roughly four hours. If there can be new ways to develop treatments for glaucoma based on THC research, this could be a new area of research and innovation for those suffering from Glaucoma.

- Low Appetite

THC leads to an increase in appetite, commonly known as "the munchies." This increase in appetite can be extremely beneficial for those with eating disorders or people who are underweight for any reason. This is also beneficial for those who are going through cancer treatments and who have a reduced appetite as a result.

- Migraines

The pain relief effects of THC are beneficial at assisting people who suffer from migraines, as many over-the-counter pain relief treatments prove to be ineffective at treating migraines.

- Inflammation

Similar to CBD and how it acts in the body, THC is a very effective way of reducing inflammation throughout the body.

Cannabidiol or CBD

CBD stands for cannabidiol. CBD is found in both hemp and marijuana, while THC is not found in significant amounts in hemp. CBD provides the consumer with much different effects than THC does. There are many ways that a person can consume CBD without consuming THC if they do not want to get high. CBD is known as a "nonpsychoactive" compound, which means that it does not give the consumer any psychoactive effects when consumed. CBD can be consumed in large amounts with little to no side-effects, as it is **not** a psychoactive compound.

The Benefits of CBD

CBD is known to help with the following, without leading to any sort of feeling of being "high;"

- Anxiety and Depression

CBD is known to help treat anxiety and depression. It can help a person to feel relaxed and at ease, and this can be great for treating anxiety. It can also help to treat depression by leading to feelings of being at peace with one's current situation. It does this by binding to receptors in the brain, which are related to *serotonin,* the brain chemical responsible for depression as it is related to mood. CBD's ability to act on the serotonin receptors in the brain makes it a great natural alternative to regular pharmaceutical drugs. One of the benefits of using CBD as a treatment for anxiety and depression is that you will be able to treat these illnesses without the side-effects that come along with pharmaceutical drugs that are prescribed for the treatment of anxiety and depression. Further, some anxiety and depression medications can be addictive and are, therefore, dangerous due to the possibility of developing substance abuse or dependency.

There are studies that have been done on children with post-traumatic stress disorder (or PTSD) wherein they were given doses of CBD to help with the anxiety associated with this disorder. The results of this study showed that CBD proved useful as a treatment for the anxiety associated with PTSD in children.

Additionally, studies have shown that CBD acts very similarly to antidepressants in the brain. There have been studies on both humans and animals, and in both samples, CBD proved to be a very successful antidepressant.

- Psychosis or Other Mental Health Conditions

Since CBD does not have any psychoactive effects, it is a great choice for those looking to deal with psychosis or any other mental health condition of this sort, such as schizophrenia. The relaxation and calm that CBD can provide a person with is beneficial in treating or improving psychosis. This is not true for marijuana, which also contains significant amounts of THC. THC will not reduce psychosis; it will actually increase it. It is important to be aware of the THC content of your product before using it to treat psychosis.

- Insomnia

CBD has been shown to be an effective treatment for insomnia in both children and adults. In these studies, CBD was useful for not only falling asleep but staying asleep as well. The reason for this is still unknown; however, the results are proven. Some hypothesize that it is the ability of CBD to induce relaxation and feelings of calm in those who consume it, which can lead to better and more restful sleep. This is because there are underlying causes for insomnia related to the mind and anxiety, so treating these underlying causes is the real way to treat insomnia.

One thing to note about using CBD for insomnia is that it usually will take some time for the results to begin taking action. In these studies, it took somewhere around four weeks of regular consumption for patients to begin noticing the effects of CBD on their insomnia.

- It is Neuroprotective

CBD has been shown to have neuroprotective effects in the brain. The word neuroprotective comes from the two words, *neuron* and *protection*. Neurons are cells in the brain that send signals, which creates our thoughts, our feelings, our movements, and virtually everything else that the body does. This all begins in the brain.

CBD has been shown to protect the structure of neurons in the brain, which prevents the loss and damage of neurons over the span of a person's life due to injury and/ or aging. This is not only beneficial for people who wish to keep their brains healthy and strong but especially for those who suffer from neurological disorders such as multiple sclerosis (MS) and epilepsy. These disorders are related to neuronal dysfunction, and CBD has been shown to help reduce the symptoms and the severity of these diseases, such as a reduction of seizures. The research in this area is still in its infancy, but the results are promising.

- Addiction Treatment

CBD is currently being studied as a treatment for addiction and substance abuse. CBD acts on the brain in areas specifically related to addiction and substance abuse. This means that CBD is an effective treatment for reducing substance dependence and the behaviors and thoughts associated with addiction. This is being heavily studied, and there is still much more information to come from these studies.

- Inflammation

CBD has been shown to reduce inflammation in the brain and in other areas of the body. One such example is the role that CBD can play in reducing the severity and the progression of Alzheimer's disease. This is because of its ability to reduce the inflammation in the brain that is associated with Alzheimer's as well as the degeneration of neurons in the brain that is part of dementia and Alzheimer's. This can help the person's condition to remain more stable and can slow the decline that comes with these diseases.

- Pain

Marijuana in various forms has been used as a treatment for pain for centuries since the first demonstrated uses of the plant in 2900 BC. One of the reasons why marijuana is so great for pain reduction is because of its CBD content. This is due to the way that CBD acts in the brain and spinal cord, where it binds to specific receptors that are associated with inflammation and sensations. CBD relieves pain in part by acting on the pain signals which travel from the site of pain to the brain, reducing your awareness of your pain- which is a large part of feeling pain. It does this by getting in the way of the pain signals as they try to reach their target sites within the brain.

Since CBD also has relaxing and mood-boosting effects, it can help a person in pain to take their mind off of the bad and switch its focus to the positive, which can help to relieve pain as it is not in the front of the person's mind.

There have been numerous studies that showed that CBD was an effective treatment for arthritis and Multiple Sclerosis. CBD is also used as a treatment for different types of pain, though most studies have been done on individuals experiencing different forms of chronic pain such as arthritis.

- Inflammatory Bowel Disease

As you now know, CBD is an effective treatment for inflammation. For this reason, it is a great way to treat disorders associated with inflammation of different areas of the body, such as the bowels. Inflammatory Bowel Disease or IBD is a disease that can lead to pain and bloating as a result of an inflamed bowel. Irritable Bowel Syndrome or IBS is another example of this type of disorder. CBD is an effective treatment for these disorders as it reduces general inflammation in the body.

- Nausea

Nausea is shown to be relieved by the consumption of CBD. The mechanism by which this works is still widely unknown, but through numerous studies, it has shown to reduce both nausea and vomiting. In this way, it can be helpful for morning sickness, side-effects of cancer treatments, and just general nausea as well.

- Migraines

Since CBD is so effective for pain relief, this is a great option for those who suffer from migraines. These intensely painful types of headaches involve tense muscles of the neck and often also bring about feelings of nausea and sometimes even vomiting. CBD will not only help to reduce the pain of a migraine, but it will also reduce nausea and vomiting associated with migraines.

- Seizures

Since the brain functions by sending small electrical impulses back and forth along its cells, the malfunction of these processes can cause problems of a variety of types—for example, seizures. Seizures happen due to a sort of "electrical storm" within the cells of the brain. This causes the symptoms of a seizure, such as convulsions, abnormal behaving, and sometimes the loss of consciousness.

Since CBD has proven to be neuroprotective- it protects the brain cells from damage and degeneration, it can help to prevent seizures in people with epilepsy, but it can also reduce the risk of seizures in people who do not have epilepsy but run the risk of having a seizure for other reasons.

- Anti-Acne

Acne is a skin condition that affects many, many people. This condition is associated with the production of an oil called *sebum* in the skin, as well as inflammation in the deeper layers of the skin. Since CBD reduces inflammation, it is an effective treatment for acne. Further, CBD has been shown to reduce the production of sebum in the skin, which further reduces acne. These studies used CBD in the form of CBD oil, which is made from the extraction of CBD from the hemp plant and mixing it with another natural oil such as hemp seed oil or coconut oil.

- Cancer Symptoms

There are a number of symptoms related to cancer and cancer treatments. These symptoms can include nausea, vomiting, pain, appetite loss, and so on.

One study gave a combination of THC and CBD to cancer patients who had not been responding to regular pain medications, and this proved to reduce pain when compared to people who were given THC-only compounds.

It has also proven to reduce nausea and vomiting that is associated with chemotherapy and other cancer treatments. In some people, regular cancer medications are ineffective, and they seek alternatives to these. CBD and combinations of THC and CBD are effective alternatives to these drugs in many cases.

- Anti-Tumor

CBD has been shown through studies to be an anti-tumor agent. In studies done on animal test subjects, CBD was proven to prevent the spread of several cancers due to its anti-tumor effects. In one study, CBD was shown to induce cell death in tumor cells in human women subjects with breast cancer. In mice, there was a study that tested a CBD concentrate in mice with breast cancer tumors, and this study resulted in the inhibition of the spread of these cancer cells in the mouse subjects. There are more studies being done to test the effectiveness of this type of treatment in human subjects.

- Improves Heart Health

CBD has been shown to reduce blood pressure in patients with high blood pressure. In turn, this reduces a person's risk of diseases such as stroke, metabolic syndrome, and heart attacks as these conditions are associated with higher blood pressure.

CBD also helps to manage and reduce the symptoms of heart disease as it reduces the inflammation of the heart and can reduce the instances of heart cell death/ heart damage that are present in heart disease. This is also thanks to its ability to lower stress levels and antioxidant properties.

- Diabetes

Diabetes is the misregulation of blood sugar due to problems with the hormone insulin. Treatment with CBD showed reductions in inflammation associated with diabetes, improving the associated symptoms, as well as reducing the incidence of diabetes in mouse subjects.

The Differences and Similarities Between THC and CBD

Below you can see a chart that will help you to remember the differences and similarities between THC and CBD.

	CBD	THC
What kind of cannabis plant does it come from?	Hemp and Marijuana	Marijuana
Illegal in Most Places	No	Yes
Gives the consumer a high	No	Yes
Side-Effects	No	Yes
Psychoactive?	No	Yes
Therapeutic Effects	Yes	Yes
Reduces Psychosis	Yes	No

The Mechanisms for Using Marijuana

In this section, we are going to break the science down to the method of consumption, as the method by which you consume marijuana determines how your body processes it and how it leads to the feeling of your high. We will look at smoking or inhalation, vaporizing, and edibles.

When growing your own marijuana plants, you will want to have the method of consumption in mind, as this could inform which strains you choose to grow, for example.

Inhaling Through Smoking

We are going to begin by looking at the mechanism by which marijuana acts when it is inhaled into the lungs or smoked. When you smoke marijuana, the THC that we just learned about it absorbed into the bloodstream through the thin layers in your lungs. Once THC enters your bloodstream in this way, it begins to give you the therapeutic effects that you seek. It does this by making its way through your bloodstream to your brain. Once the THC gets to your bloodstream, it will make its way to your brain in just seconds.

Smoking marijuana is the quickest way to feel the effects of a marijuana-induced high. It is also the most common way to consume marijuana. Within the lungs, there are millions of small pockets of air. This is where the exchange of air occurs whenever your inhale- the clean air goes into the bloodstream, and the used air goes into the lungs to be exhaled into the atmosphere. When you inhale air that is filled with marijuana smoke, the gas exchange that occurs moves the air containing THC and CBD into the bloodstream through these small pockets of air called *alveoli*.

This is how THC makes its way into your system. The reason that inhaling marijuana is the quickest way to get high is that these small pockets of air are plentiful, and they are formed by very thin layers of membrane. This means that the exchange of gas across these membranes happens very quickly. The other reason is that these alveoli lead the consumer to feel such a quick high is because these alveoli have such a high surface area, and this surface area allows the lungs to absorb a lot of air in a short amount of time. Then, once this THC makes its way into the bloodstream with ease, it also quickly makes its way to the brain as it joins the regular stream of blood that is pumping its way through your body. Since this happens all day every day without your knowledge, your body is extremely efficient at processing the air you breathe in and passing it all around your body. When you add THC into this blood flow, the body is still just as efficient at passing it around.

Once the THC reaches the brain- seconds after it reaches the bloodstream, it begins changing the chemistry of the brain. The most common initial feelings that marijuana use elicits are feelings of relaxation and light-headedness. This is the feeling that many people enjoy and is what keeps them coming back to marijuana. After some time, the effects of marijuana consumption can change from relaxation to paranoia and anxiety.

The effects that a person feels and the extent to which they feel them is largely dependent on the strain of marijuana that they consume (We will discuss this more in-depth later), the amount they consume, the way they consume it and finally, the person themselves and their own personal brain chemistry. In general, THC acts in the brain by impersonating the brain's own chemicals, thereby interfering with the regular operations of the brain.

There are specific spots within the brain where THC and CBD molecules can attach themselves. The areas where these spots are located are responsible for learning, problem-solving, coordination, and short-term memory. This may explain why you feel that these areas of your brain are affected when you smoke. For example, THC acts on the hippocampus, which is an area of the brain associated with short-term memory. When THC binds to the receptors in the hippocampus, this makes it difficult for the brain to make new short-term memories. Another example of this is the cerebellum, which is an area of the brain that is responsible for balance and coordination. When THC binds to the receptors in this area of the brain, it makes it difficult for the cerebellum to control the coordination and balance of the body.

Inhaling Through Vaporizing

When using a vaporizer to inhale marijuana, the marijuana becomes heated enough for the active chemicals in it to become vapor, and when your inhale this vapor, it works in the exact same way as smoking does once in your lungs. The main difference is that some people prefer to inhale vapor rather than smoke. When you use a vaporizer, the marijuana isn't heated enough to burn it, and this is why you inhale vapor instead of smoke.

Consuming Through Edibles

When consuming marijuana in the form of edibles, you ingest it just like any other food. This is usually done by baking oil made from the marijuana plant into something like a cookie or drinking it in tea. Once you eat the edibles, they enter your digestive system through your mouth and make their way to your stomach. Once they enter your stomach, the THC is absorbed into your bloodstream through the walls of your digestive system. Once in your bloodstream, the THC makes its way to your brain in the same way that it does in the above methods, leading to the feeling of being high. The difference between ingesting marijuana through edibles and inhaling it through smoking is in the absorption of THC.

This is the part where you can see clear differences and can form preferences when it comes to your methods of choice. As I mentioned, when you inhale marijuana through smoke, the absorption of THC is very quick because of the large surface area and thin membranes of the alveoli. When you ingest it by way of edibles, however, your body absorbs the THC much slower. This is because the stomach is much slower at absorbing. The membranes are thicker, and there is less surface area for absorption. Because of this, however, one of the reasons why some people choose edibles over inhalation is because while the effects of the high take longer to come on, they last much longer than they would through inhalation.

CHAPTER 2: UNDERSTANDING CANNABIS PLANTS

When it comes to Marijuana, there are three main types that people normally choose from. For a novice marijuana user, it may not matter all that much which type of marijuana they choose to use. However, for frequent marijuana users or those who use it for medical purposes, it matters a whole lot more which type of marijuana they choose to use. For this reason, it is important for those who are considering growing their own marijuana plants to understand the differences that exist between different marijuana strains so that you can provide yourself or your customers with the right product.

In this chapter, we will be learning about the three different types of marijuana; Indica, Sativa, and hybrid. We will also learn about the cannabis plant in more detail, including how to tell the sex of your plant and what this means for your garden.

Indica, Sativa and Hybrid

As I alluded to in the introduction to this chapter, cannabis is a plant that can be divided into three subspecies; Indica, Sativa, and hybrid.

Indica and Sativa plants are different in the physiological effects that they will provide the user with, as well as in their outward physical appearance. The main differences between indicas and sativas are their medical effects and how they affect the users' productivity levels and energy levels.

The effects of the Indica strain of marijuana include what is known as more of a "body high," which includes relaxation, appetite stimulation, sleep aid, and pain relief. Indicas are recommended for nighttime use as they give the user heavy relaxation and help to fall asleep. Indicas tend to lower a person's energy and are best for nighttime consumption and is often used to help relax after a full day of work and activities. Strong Indica strains are known to give its users 'couch lock' which is a slang term of feeling so relaxed that the user can barely get up from where they are sitting (usually the couch). The Sativa strain of marijuana plants provides the user with more of what is called a "head high," which includes alertness, euphoric and uplifted feelings, creativity, and increased energy. This strain is best recommended for daytime use as it does not give the same tiredness effects that indicas are known for. Sativas are known to be cerebral and uplifting that helps a person enhance their productivity and even their creativity. While indicas give a person a 'body high' (hence the couch lock), sativas bring more of a 'mind high.'

Those who use marijuana to treat chronic pain normally will select an indica as it is known for their pain-killing benefits. However, others who may need to go to work or have family responsibilities may choose sativas as they still require lots of energy to get through their day. When it comes to specific ailments, sativas are much better for psychological disorders such as anxiety, PTSD, and depression. Indicas are chosen more often for disorders like inflammation and pain; hence it helps those with cancer, fibromyalgia, and arthritis. However, since so many ailments and diseases are accompanied by psychological disorders like insomnia and depression, a person must consider choosing a strain that can treat their core disease while being able to manage other symptoms that come with it.

A Sativa and an Indica can actually be bred and mixed together to create a hybrid strain. Indica and Sativa mixes are very common and are known for their 'alert mellowness,' which still allows for productivity.

Hybrids can be bred by mixing one Sativa with one Indica parent or using two Sativa parents or two indica parents. Since there are so many hybrid strains available to us, many can be bred to possess abilities to kill pain whilst not giving the user couch lock during the day.

Those who use marijuana to medicate during the daytime will often use a Sativa-dominant hybrid that will allow them to function during the day but will switch to an Indica-dominant strain in the evenings for pain relief and relaxation. Most hybrid strains that are available through dispensaries are labeled as either 'Sativa-dom' or 'indica-dom.' This is quite self-explanatory as it simply tells you which strain is more dominant. Depending on the dispensary you are purchasing from, strains can be as specific as '60/40' Sativa/Indica or '80/30' Indica/Sativa.

Common Strains and Their Characteristics

Over the past few decades, there have been over one thousand different marijuana strains that have been bred. It is important to understand what the different types are if you are someone that is using marijuana to treat a certain problem. Some types of marijuana are more appropriate to treat certain ailments and diseases, while others may not have any effect at all.

Being able to choose the right strain is extremely important when ensuring that your problems can be solved. Like I mentioned above, however, if you are a recreational user that is new to marijuana, strains matter much less to you.

In the modern-day, professional marijuana growers will breed a wide variety of strains in both Indica and Sativa types solely for those who use it to medicate any ailments. This has offered the market a much larger variety of marijuana where one can choose the best strain to match their preference, lifestyle, or disease.

In this subchapter, I will be teaching you about a few different types of strains and what effects they have to offer. This will help you better determine which strains you'd like to try and what they have to offer, as well as which strains you might like to begin with as you grow your own marijuana plants.

Indica 1: Granddaddy Purple

Granddaddy Purple or otherwise known as GDP, is one of the most famous indicas out there. It's a California staple that is a mix between Purple Urkle and Big Bud. This strain has a complex berry and grape aroma that comes from the Purple Urkle. The oversized bug structure comes from the Big Bud strain.

The flowers of Granddaddy Purple are in a shade of deep purple with a snow-like dusting that is white crystal resin. Its effects are potent and gives the user both a body and mind high that feels like physical relaxation and euphoria at the same time.

Your thoughts may feel as if it's in a dreamy state while your body feels fixed into one spot for the duration of the high. GDP is normally used to help those fight muscle spasms, appetite loss, insomnia, stress, and pain.

Sativa 1: Sour Diesel

Sour Diesel is a strain that is named after its diesel-like smell. It is an invigorating strain that is sometimes known as Sour D. This strain is fast-acting and gives the users a dreamy and energizing high that built the strong reputation it has today. This strain is also great for healing depression, pain, and stress and actually has quite long-lasting effects. These effects make Sour Diesel a popular choice for those who medicate with marijuana. There aren't many drawbacks that come with this strain besides the occasional dry mouth.

This strain was created back in the 90s and is rumored to have been created by breeding one 50/50 hybrid strain called Chemdawg 91 and an indica-dominant hybrid called Super Skunk.

Hybrid 1: Blue Dream

Blue Dream is a hybrid that is Sativa-dominant that has achieved an incredible status from the California region. It is a cross of Haze and Blueberry, which a balance of mind invigoration and full-body relaxation.

Both experienced and novice marijuana users enjoy these effects of Blue Dream, which eases you into a euphoria, which is both calm and uplifting. The aroma of this strain is sweet and berry-like, which comes from the Blueberry parent. This strain gives the user a quick symptom relief minus the heavy sedative effects. Due to this, Blue Dream is a popular strain for those who need to use it as a medicine during the day time. This strain is great for treating nausea, depression, and pain.

Hybrid 2: Strawberry Banana

Strawberry Banana is a strain that is otherwise known as Strawana. It is an Indica dominant hybrid that was developed by DNA Genetics. This strain is a cross between Bubble Gum and Banana Kush. Its flavors are fruity and sweet, hence the name it was given. It is well known for its high-THC content and heavy resin production. This strain gives the user peaceful and happy effects that help improve one's sensory awareness and creativity. The main effects of this strain are relaxation, happiness, and euphoria. There are minimal drawbacks with this strain, which makes it great for novice users.

Identifying a Cannabis Plant

Indica and Sativa plants are different in physiological effects and physical appearance.

- Height, Width and Leaf Size

Sativa plants are taller, skinnier and appear to be more lanky, with leaves that are pointy and thin. Indica plants are stocky and short with leaves that are chunky and broad.

- Growing Time

Compared with indica plants, Sativa plants require more time to grow and yield fewer flowers. Due to this, indica plants have dominated the market for many years since the sole focus is profit.

- Smell

When it comes to the smell of the different strains, Indica tends to smell skunky, earthy, and musty, whereas sativas smell spicy, fruity, and sweet. The main difference in the smells of marijuana are due to terpenes, which are the molecules within the plant that are similar to cannabinoids.

Cannabis Plant Sex

Marijuana plants are either female or male (known as *dioecious*). This is a concept that may surprise you, but believe it or not, your cannabis plant will have a sex- either male or female. This cannot be determined right away, but instead is able to be seen during the fifth stage of cannabis plant growth- the flowering stage. We will discuss the different stages of cannabis plant growth later on in this book so that you have a better idea of when to begin looking for this on your own plants. At the flowering stage, you will be able to begin determining the sex of your plants.

The way that male and female plants procreate is similar to the way in which humans will have sex to produce a child. The resulting child will be made up of half of its mother's DNA and half of its father's DNA. Two cannabis plants- one male and one female will come together for reproduction, and the male plant will pollinate the female plant, leading the flowers of the female plant to produce seeds. If your female plants are pollinated, they will die shortly thereafter, as it has then achieved its life goal of passing on its genetic material. For this reason, understanding the sex of your cannabis plants is necessary so that you can control the pollination of your plants and the lifespan of your female plant's flowers.

You would want to do this so that you can decide whether or not you want your female plants to be pollinated by your male plants.

If you are looking to attain seeds to sell or to re-plant in order to grow a new harvest, you would want your female plants to be pollinated. If you wish to harvest the flowers or the buds of your marijuana plants in order to sell, smoke, or transform them in some way, you will not want your female plants to be pollinated by the male ones.

In order to determine the sex of your plant, you will need to check to see which reproductive organs your plant is showing- male or female. This is similar to a human baby, where you must look at their reproductive organs to determine their sex once they have formed. Most people will sell only buds that have not been pollinated; therefore, they have no seeds. The buds that you can get your hands on- through your dealer or in a store are all produced from female plants, as male plants only pollinate female plants but do not grow buds themselves. Seedless buds are called *Sinsemilla*. You can determine the sex of your plant by looking at the spaces between the nodes of your plant (the area where the plant branches off from the main stem).

Similar to male humans, male cannabis plants will have two ball-shaped pollen sacs contained at their nodes. These pollen sacs contain their genetic material, which they would then pass on by pollinating a female plant. Female cannabis plants will have something called a *Stigma,* which is a small hair-like structure that is used to catch the pollen that the male plant produces- this is how they essentially "have sex" and reproduce.

The reproductive organs on a female plant will begin growing somewhere around the fourth week of the plant's growth cycle, but for some plants, it may take around six weeks instead. Patience is key in this stage. When you can begin determining the sex of your plants, their sex organs are not fully developed, so they are not actually able to serve their purpose yet, which gives you time to determine which of your plants are males and remove them before pollination can occur.

CHAPTER 3: GROWING MARIJUANA

In this chapter, we are going to look at how you can begin to grow your own marijuana plants. You know have a solid foundation of knowledge regarding cannabis and marijuana plants, and now we are going to take it up a notch and look at what it takes to grow marijuana. In this chapter, we are going to look at the different options available to you when it comes to growing your own marijuana plants and what to consider before choosing your method.

Since cannabis is classified as a type of weed, it has incredibly high resilience. Growing marijuana indoors isn't as hard as it may seem. As long as you have a basic knowledge of plants and the ability to follow a set of instructions, this task shouldn't be too difficult for anybody to do. If this is your first garden, you are lucky that you are growing Marijuana, as it is a classic "weed" in that it can withstand a wide variety of conditions. If you make mistakes, this plant will be more forgiving than some other plants that you could have chosen to grow. The important part about learning to grow cannabis indoors is making sure that you understand the stages of cannabis growth and working with varieties that are easy for beginners to grow.

Options for Growing Marijuana

Typically, there are two methods that growers use to grow their marijuana plants. The first is the traditional method, growing marijuana in soil. The second is a new-age method, which involves something called hydroponics. I will walk you through both of these methods in this chapter, along with how these different methods each come with their own set of benefits and drawbacks. There is also one other thing to consider, which is whether you are going to grow your marijuana plants inside or outside. Throughout this subchapter, I will outline the different options available to you so that you can make the best choice for your marijuana garden.

Growing Marijuana Inside Versus Outside

Cannabis has been grown outdoors for the majority of its existence since it is a plant that was discovered naturally growing in large fields in a variety of places around the world. It is only in recent times that marijuana has been grown indoors. This change from outdoors to indoors came about mostly out of necessity. This necessity came from the strict laws about marijuana that required growers to keep their cannabis plants out of sight so that they would not be caught growing them.

Now that there are more sophisticated ways to grow cannabis indoors, it is a method of choice by many. The first decision that you need to make when beginning to grow your own marijuana plant is whether you will grow your marijuana plants inside or outside. In order to make this decision, there are a few things that you will need to keep in mind.

1. Climate

You will need to think about the climate in which you live when considering how you will grow your marijuana plant. If you live in a climate that is variable, and that involves a lot of drastic temperature changes, growing marijuana outdoors will likely prove difficult for you. If you live somewhere consistent in terms of temperature and precipitation, you will have more choice about whether you wish to grow your plants inside or outside.

2. Space Allocation

If you choose to grow marijuana outside, you will be freer to allow your plants to grow freely without worrying about the space that they will take up. If you grow your plants indoors, you will need to ensure that you have enough space allocated for your plant to grow, keeping in mind that it could grow quite tall, depending on which particular strain you choose to grow.

If you do not have the space to dedicate to your marijuana plant indoors, growing it outdoors will be a better option for you.

3. Control Over The Environment

If you choose to grow your marijuana plant indoors, you will have complete control over every aspect of the environment that your plants are growing in, including the type and amount of light, temperature, water, air, and so on. If you grow outdoors, it is up to the environment in which you live and the time of year that you are growing, since this will affect things like the amount of sunlight present, the temperature and the amount of water the plants are getting through precipitation.

When grown indoors, the marijuana plants that result tend to have higher amounts of THC, as well as a more aesthetically pleasing look. This makes them great for selling if this is your goal. On the other hand, plants grown outside are able to receive their light from the sun (instead of from lights trying to emulate the sun), which results in higher yields and stronger plants. This is assuming that the climate outside is suitable for the growth of marijuana.

Now that you understand some of the factors that you need to take into account when deciding whether to grow your cannabis plant indoors or outdoors, we will look more specifically at each of these two options, as well as what options you have within each of these two locations- growing marijuana using soil or using hydroponics.

Growing Marijuana Using Soil

Growing your cannabis plant in soil is the most traditional medium for growth, both indoors and outdoors. Growing your plant in soil is also the most forgiving method of growth, which makes it the best choice if you are a beginner grower.

Pretty much any high-quality soil will work for growing cannabis, as long as the soil doesn't have artificially extended realize fertilizer contained within it (such as Miracle Gro), which is not ideal for growing high-quality cannabis.

Growing Marijuana Using Hydroponics

Hydroponics is a method of growing plants that do not require using soil. Instead, the roots of a plant are grown in clay pellets, rockwool, coco peat, water, gravel, or sand. The nutrients required for a plant are mixed in a nutrient solution that is given to the roots directly.

Any water that is left over and wasn't absorbed by the plant will be recycled through the system to be used later.

In the modern day, many indoor cannabis growers are turning to this new, soilless hydroponic growing method. This method requires the grower to feed the plant with a liquid nutrient solution that is concentrated with minerals, salt, and nutrients which the plant needs to grow. The plant will absorb these nutrients directly into its roots using the process of osmosis absorption.

Growing your cannabis plant using this method allows for quicker nutrient uptake and can lead to bigger yields and faster growth than using traditional soil growth. However, it requires the grower to have higher precision and skill as plants will respond quite quickly if they are under or overfed. They are also more susceptible to lockout and nutrient burn. We will discuss these different options later on in this book so that you can learn how to prevent and deal with these different potential struggles.

Soil-Based Gardening Versus Hydroponics

In this subchapter, we are going to look at some of the similarities and differences between soil-based gardening and hydroponics.

To begin, the major difference between hydroponics and soil-based gardening is that hydroponic systems grow plants without the use of soil and by giving the plants chemical nutrients instead of natural nutrients from the soil in which they are grown. Further, hydroponics uses artificial light sources, whereas soil-based gardening most often uses natural sunlight.

In a hydroponic system, the plants are able to take up water into their roots directly from the water in which they are growing. In soil-based gardening, the plants must wait until water trickles down through the soil to their roots, and thus, less water is available to them. It is for this reason that hydroponic systems require less water because the water that the plants are suspended in also acts as a means of feeding them.

When planted in soil, plants will spread their roots as far as possible in the soil in order to get access to as much water as possible. This means that the plants must be grown apart from each other to enable each plant to spread its roots enough. In hydroponic systems, the plants do not need to spread their roots out as much because they have unlimited access to water, so plants can be grown much closer together which will result in a higher yield per area, or a much smaller space needed to grow the same amount of plants.

Another difference between these two methods is that in hydroponics, plants are given constant access to water, nutrients, and oxygen, which means that they are not reliant on rainfall, sunlight cycles, or fertilizer. For this reason, plants grown in hydroponic systems are able to grow much faster than those in soil-based gardens as there is much less competition for limited resources.

There is no one method that is better than another, only the method that is right for you and your purposes. By learning as much as you can about the similarities and differences between different types of gardening systems, you will be able to determine what works best for your personal gardening goals.

CHAPTER 4: THE CANNABIS PLANT LIFE CYCLE

In this chapter, we are going to learn about the cannabis plant itself and how it grows, so that you can learn what to expect when you begin to grow your own cannabis plant. The cannabis plant requires certain elements in order for it to grow and bloom into the beautiful buds and flowers that will eventually bring you pleasure and relaxation in the form of CBD and THC.

Throughout this chapter and the next, we will look at the stages that a marijuana plant goes through as it matures from seedling to fully matured plant, as well as how you can grow the healthiest possible marijuana plant by providing it with the right amounts of water, sunlight and food or nutrients. This chapter will help you to learn how to get the most from your marijuana seeds, no matter what your purposes may be.

There are a few things to remember about cannabis plants before we begin this chapter. The first thing is that cannabis plants can be either male or female. The second thing is that growing a cannabis plant will take somewhere between four and eight months.

You must be patient when growing a plant as it takes time and care to get it right. This chapter will walk you through the steps that you will need to take as you watch your plant grow.

Stage 1: The Seed Stage

The first stage in the cannabis plant life cycle is the seed stage. This stage is what you will be starting off with before you plant anything and before anything has grown. This stage involves a humble seed and a few steps that you will take in order to help it begin to transform.

There are a variety of ways that you can get your hands on marijuana seeds that you will then plant in order to grow a marijuana plant.

The seeds of a cannabis plant come from another cannabis plant. These seeds are the plant's way of passing on its genetic material to make progeny plants, just like humans do with their sperm or eggs. The seeds of the plant contain information about the mother and the father of the plant, just like humans.

As you learned, marijuana plants are either female or male (this is known as *dioecious*). This is similar to the way in which humans will have sex to produce a child. The resulting child will be made up of half of its mother's DNA and half of its father's DNA. The same can be said for a plant. Two cannabis plants- one male and one female, will come together for reproduction, and the male plant will pollinate the female plant, leading the flowers of the female plant to produce seeds.

After the flower of the female plant has been pollinated by a male plant, the female plant will start dying, because then it will have achieved its life goal of passing on its genetic material to possible progeny.

When the female plant is pollinated, one of the following two things will happen to the female's seeds;

1. They will fall to the soil on the ground and begin the next stage of the plant life cycle, which involves the growth of a new plant that will be made from the original female and male plants' genetic material. The new plants that grow from these seeds can be viewed as the children of the original parent plants.

    Or

2. The seeds will be harvested by farmers or by seed suppliers in order to be turned into oils, food products, or to be sold as seeds for others to plant. The farmer themselves may also plant the seeds in their desired locations in order to grow a new crop of plants.

Because of this, you will need to find someone to collect the seeds from their pollinated female marijuana plants, or you will need to buy them from a marijuana seed supplier, in order to begin your own marijuana plant crop. In addition to this, you will eventually need to decide if you wish to have your own female marijuana plants pollinated so that they will produce seeds for you, or if you instead wish to have them flower for you so that you can use this to smoke using a bong or joints. We will discuss this second point later on in this book, but for this stage, it is important to understand where your seeds come from and how to get your hands on them.

When you attain your marijuana seeds, you can determine their quality by the look and feel of them. They should be brown in color- either a light or a dark brown. They should also be dry and hard to the touch.

Your seed must be mature enough in order to grow into a plant. If the seeds are white or green in color and soft to the touch, they will not be able to grow into a plant as they have not adequately matured.

Once you plant the seed, it requires some attention in order to bring it to life. It is not ready to begin growing until it receives some attention from you first. What it needs most at this stage is water. We will get into this in more detail in stage 2, as it will become important for reaching the next stages of the cannabis plant life cycle.

What you should focus on most at this stage is ensuring that before you begin to germinate or grow your seeds, you have made sure that you will have enough time and space to care for your plant and allow it to grow freely. This will help promote its success. What this means is that you want to ensure you have the time available in your life to care for your plant as it grows, as this will be necessary for success.

You also want to ensure that you have the physical space available for the plant to grow and mature without being restricted. This space can be either outside or indoors. We will discuss this in more detail in chapter six.

Whichever location you choose for growing your marijuana plant, you must still ensure that you have enough time and space for it to grow into the best possible product.

When to Plant Your Seeds

The best time to plant your cannabis seeds is in the months of March and April. These months are best for your plant seed to germinate, which you will learn about in the next section.

If you are new to marijuana plant growing, planting your seeds during these months will help you to find the most success possible.

Stage 2: Seed Germination

The next stage of the cannabis life cycle is called Germination. This stage involves the first stages of change in the seed, and it begins the growth of your plant. This stage will take somewhere between five and ten days.

Some people refer to this stage as "the popping stage" as it involves the "popping" of the seed. As I mentioned, this stage is encouraged by watering the planted seed. The seed may also require heat and air in order to encourage the popping of the seed.

These three components, heat, water, and air- are what will lead your plant to pop in just the right way. I will outline one of the most common methods for achieving cannabis seed germination below so that you can follow along and try it for yourself.

1. To begin germinating your seeds, you will need to have access to paper towels and distilled water. Begin with three to five pieces of paper towel and put them together in some distilled water, enough for them to soak. When you remove them from the water, let the excess water drip off of them and don't proceed to the next step until they have finished dripping and are soaked but not spilling over with water.

2. Take two pieces of soaked paper towels and put them down on a surface like a plate or a dish. Take your cannabis seeds and place them on top of the soaked paper towel, leaving enough room between each seed to make room for the seed to open up- about an inch or so. Put the rest of your paper towel pieces on top of your seeds on the dish so that they are now sandwiched between two layers of soaked paper towel.

3. Take another plate, dish, or bowl and flip it upside-down on top of the seeds covered with paper towels soaked in distilled water. This creates a moist, dark, and warm space in which the seeds can comfortably begin to germinate.

4. Keep this contraption somewhere that is warm enough for the seeds to stay humid (preferably between 69- and 89-degrees Fahrenheit).

After you have done these steps, you must wait for your seeds to germinate. While you do this, you can occasionally check to ensure that the paper towels are still moist enough. If they are beginning to dry out, saturate them again with water to maintain an ideal germinating environment. This environment will include heat, water, and air.

You will know that your seeds have reached germination when they have "popped" or split open. When this happens, you will see a small sprout that has begun to grow from this split area of the seed. The amount of time that this takes will depend on the seed, so be patient with your little seeds.

One important thing to note is that when your seed splits and the sprout can be seen, it is important to keep this area free of contamination.

One of the most common contaminants is your hands. Keep your hands away from the sprout, as this is the part of the plant that will become the main stem of your plant eventually, and you must refrain from interfering with it. This part of the germinated seed is called the *Taproot*.

Once your seeds have germinated, you are ready to plant them in soil. This is when you are ready to begin growing your plant.

In order to plant your germinated seed in soil, you will begin with a small pot and some lightly packed soil. Using a small cylindrical device like a pen, poke a hole in the soil. This will be where you will insert your seed.

To avoid contaminating your taproot, pick up your seed using chopsticks or tweezers and transfer your popped seed into the small pot of soil. When you place it into the soil, ensure that the taproot is facing down. This may seem counter-intuitive, but it is the correct way to plant your germinated seed. When you put the taproot into the hole, push the seed down into the soil about one-quarter of an inch. Cover the exposed seed and the top of the hole with some of the soil.

Once you have planted your germinated seed, you will need to begin watering it. You can do this using a small spray bottle so that you do not over-water your plant.

You will need to provide the seed with just enough water to grow, but not so much that the seed cannot get any air. Growing a plant is a delicate balance of all of the elements it needs. You also need to ensure that your seed stays warm enough. Keep it in a warm environment, away from any windows if you live in a cold climate. Choose somewhere warm and water the seed lightly so that the air and the water are both taken care of. At this stage, it needs to be in the dark, and the darkness factor is already taken care of for you since the seed is submerged in the soil. Your seedling will begin to show through the soil in about a week if everything goes as planned.

You must remember that when waiting for your seed to sprout, there are many factors at play. This stage requires patience and an understanding that seeds are very delicate. Further, every seed is different, and you must remember that not all seeds will respond or behave in the same ways. Some may take longer to sprout than others, and some may not sprout at all. Some seeds are not able to sprout for one reason and another, so planting many seeds at once is the best option. If some have sprouted but not others, they may take a bit of extra time and attention.

By planting several, you can ensure that at least one or two will sprout as expected and can continue to grow into plants. As your plant begins to grow out through the soil, you will begin to see two small leaves form on the stem. These initial leaves are a great sign, as they are what will be used to capture sunlight and use it for the plant's nutrients and health. At this stage, the plant will have grown out of the casing of the seed and will be growing on its own. The roots of the plant will be growing and spreading out beneath the soil, in an effort to anchor the plant in the soil in preparation for it to grow bigger and bigger.

Once you see the first set of "weed leaves" growing on the stem of your plant (those leaves that are internationally known to symbolize marijuana), then this is when you are able to begin calling your plant a *seedling*. This brings us to our next stage in the cannabis plant life cycle, which is the Seedling Growth stage.

Stage 3: Seedling Growth

This stage of the cannabis plant life cycle can take anywhere from two to three weeks, depending on the seedlings. The first classic "weed leaves" that you see growing will only have one leaf per set, instead of the traditional 5 or 7 leaf sets that you are used to seeing on cannabis plants.

As more leaves grow, you will begin to see the leaves becoming more and more like those traditional marijuana leaves as the plant matures. The classic marijuana leaves will have somewhere around five and seven leaves on each set. Your plant should have leaves that are bright green in color; this is what indicates a healthy and thriving plant.

One very important thing to keep in mind is that at this stage, you must not over-water your seedling. Underneath the surface of the soil, the roots are still young and have not fully developed yet. They still need air, and giving them too much water can lead them to suffocation as they will not have enough room for air in this case. A small amount of water on a regular basis will be enough to keep your seedling healthy and vibrant. Another reason that it is important not to over-water your plant at this stage is that it is very susceptible to mold. Having excess moisture in the soil and around the plant can lead it to develop mold. In this susceptible stage, it is also able to easily catch a number of plant diseases, which can compromise the further growth and prosperity of your plant, not to mention the level of confidence you can have when ingesting the resulting flowers. Your plant will be considered a seedling until it begins to grow leaves that have the traditional number of leaves per set. When each new set of leaves it grows contains this number of leaves per set, this is when it has progressed to the next stage.

Stage 4: The Vegetative Stage

Once your plant is no longer considered a seedling (when it begins to grow traditional marijuana leaves with 5-7 leaves per set), it moves onto the next stage, which is called the Vegetative stage. This stage can vary greatly in length, lasting anywhere from three to sixteen weeks in length. The amount of time that this stage lasts can depend on a number of factors, which we will examine here. During the Vegetative state, the plant will begin to grow much bigger and faster than it has yet. When your plant has reached this stage, you will be ready to transfer it into a bigger pot. This will allow it to grow freely and without restriction. The roots will begin to grow and mature quickly as well as the leaves.

Vegetative: Training and Topping

At this stage, you can begin doing two things to your plant; *Training* it and *Topping* it. We will look at what these two terms mean here. Training your plant is something that is done in order to maximize your possible yield and the potency of the buds that your plants create. The practice of training your plant involves intentionally changing the chemical balances by interfering with the growth of the plant so that it has to adapt to these changes, and thus, you are left with a stronger and more resilient plant.

This is because when cannabis plants grow on their own in the wild, they want to reach their top bud as high as they can. When you are growing cannabis for commercial purposes, however, you do not want just one large and tall bud, but many, many buds. This is where training comes in. The natural method of growth does not allow the buds on the bottom of the plant to get as much light as they need, so by training the plant, you also allow the plant to have evenly healthy buds instead of the healthiest buds at the top of the plant and the buds at the bottom suffering more. There are to ways to train your plant: *High-Stress Training (or HST) and Low-Stress Training (or LST)*.

Low-Stress Training is a method that is best done for plants that are growing in indoor gardens and that are exposed to light sources for growth. LST can increase your yield greatly. This is a method that is usually done in the vegetative state, but that can sometimes also be done in the flowering stage. LST involves influencing the direction and method of growth that your plant wants to create naturally. For example, if you see that one branch is becoming much too tall, in order to avoid growing an uneven plant with uneven buds, you will tie this long branch down so that it does not continue to outgrow the rest of the branches.

You can also tie down the top branch of the plant so that it does not grow too big and shade the rest of your plant. This allows the chemicals running up and down the stem and branches to be spread out more evenly throughout the plant, leading to more even growth of the branches and buds and the overall plant. To accomplish this, use plant-specific tape, which is designed with the health of the plant in mind. When you properly train your plant in this way, the result is many branches that grow around the pot that the plant is contained in, and then when the buds grow, they grow at an even level to each other, able to be exposed to the right amount of light without being shaded by other areas of the plant that are covering them. One other way to accomplish this type of training is to use a screen placed above the plant that will act as a sort of barrier of growth. When a branch grows through the screen, you can bend it and send it back down into the screen so that it continues to grow in a downward fashion. One other method of LST that can be useful but that is a little more invasive is to bend the branches of the plant in certain places so that it kinks and then begins to grow in a downward fashion. This method is called *Stem Mutilation or Super Cropping*. You do not want to snap the branch; you just want to put a kink in it so that it stays bent. If you think you will snap the branch before you bend it begin rolling it between your fingers until it becomes softened, and then it should be able to kink much easier.

High-Stress Training is a method of training that is great for plants that are growing outdoors and have more space to grow. This method of training should be restricted o the vegetative stage of plant growth since it involves stressing the plant. Stressing the plant should not be done after the vegetative stage as it can interfere with the growth of the buds, which is the most important part of the plant growth if you are growing it for use as marijuana. HST involves breaking off parts of the plant at the top. To do this method, you will look for the newest location of growth at the very top of the plant and remove this portion. What this does is lead to the growth of four branches off of the top of the plant instead of only one or two, which will lead the plant to grow out in a more even manner instead of growing taller than it is wide. This method can carry risks as it leaves your plant vulnerable to infection or disease as you are removing the topmost section of the plant. This method will be the most time consuming, so ensure that you have the time to devote to it. There is one other method of High-Stress Training which is called Topping. Topping is much simpler than the previous method as it does not involve any tools and is less risky. Topping involves removing a small part of the top of the plant with your fingernails. What this does is send stress signals to the rest of the plant, which promotes growth in the lower areas of the plant.

This leads the plant to grow outward more than it grows upward, which allows the full plant to have access to sunlight instead of just the branches at the top of the plant. This method is to be repeated over and over again throughout the vegetative stage of plant growth in order to continue to promote the lower growth of the branches.

You can also *Prune* your plant in order to keep it potent and healthy. Pruning involves searching for areas of the plant that you determine will not receive adequate sunlight or nutrients because of the look of them or their location on the plant (i.e., in very shaded areas of the plant). You can remove these areas of the plant by snipping them off, which will allow the plant to redirect its energy and resources to other areas of the plant in order for them to grow and flourish. This will lead to maximum potency and the best parts of the plant being provided with what they need to grow and be strong.

At this stage, you can begin to examine your plant in order to determine different things about it. For example, you can look at the space that exists between each of the nodes on your cannabis plant. The nodes are the parts of the plant where there are new stems containing leaves that jut out from the main stem of the plant.

If the space between the nodes is small and your plant is proving to grow densely in its leaf-spacing, you are growing an Indica strain of cannabis. If the node spacing is large and the plant is growing longer and more spread out, you are growing a Sativa strain of cannabis.

As the plant grows and grows, it is necessary to begin watering your plant more and more. When it is small, the watering is primarily focused on the soil directly under the stem. As the plant grows in this stage, however, you are able to begin watering the plant at a wider circumference from the stem as the roots will have spread out underneath the soil. This will ensure that the plant stays healthy and is taken care of in a well-rounded manner. This will also allow the outermost roots to grow and develop, which will keep them spreading out and will allow the plant to grow and flourish.

At this stage, you can also begin using soil that contains some nutrients, especially nitrogen. As the plant grows and spreads, it will need more nutrients to stay strong and healthy. When the plant is young, this is not as necessary as it mostly requires water and air, but later on, in the process, it will begin needing food. This is similar to a human baby beginning to eat solid food instead of just milk as it reaches its stages of rapid growth.

Further, at this stage, the plant needs many hours of sunlight per day as it will use this light to grow and develop. It needs 18 hours of sunlight per day. In this next section, we are going to look at the stages of the cannabis plant growth that involve the flowering and the bud growth that you have been waiting for all along!

Stage 5: The Pre-Flowering Stage

The flowering stage begins, as you can imagine when the buds begin to develop. If you are growing your cannabis plant indoors or in a controlled environment, you will need to induce flowering intentionally by changing the amount of light that you allow the plant to receive. This is because when a plant is growing naturally in the wild, it will be affected by the changing of seasons as this comes with changes in the amount of sunlight in a day, the amount of water in the form of rainfall and the temperature outside. Naturally, this stage of growth will coincide with a reduction in the amount of sunlight that the plant is exposed to in a day from about 18 hours to around 12 hours. Naturally, this stage would occur at the end of the summer as the days begin to shorten.

If you are growing your cannabis plant indoors, in a controlled environment or at a time of year or in an environment when plants would not normally bloom, you will need to change the amount of light you are exposing your plants to at this stage in order to induce flowering.

Stage 6: The Flowering Stage

This is likely the stage when you will feel most rewarded as you will be able to begin seeing the fruits of your labor coming to fruition. As you see, the flowers begin to develop on your plant; you will know that your plant is reaching maturity. The buds or flowers are what marijuana plant growers are waiting for all along, so this stage will come with a lot of excitement for you.

Cannabis Plant Sex

As I mentioned at the beginning of this book, at this stage, you will be able to begin determining the sex of your plants by examining the sex organs that your plants begin showing. It is at this stage that you will decide if you will allow your female plants to be pollinated or if you will instead remove the male plants from your crop.

Male Versus Female Cannabis Plants

As I mentioned, you can determine the sex of your plant by looking at the spaces between the nodes of your plant (the area where the plant branches off from the main stem). Similar to male humans, male cannabis plants will have two ball-shaped pollen sacs contained at their nodes. These pollen sacs contain their genetic material, which they would then pass on by pollinating a female plant. Female cannabis plants will have something called a *Stigma,* which is a small hair-like structure that is used to catch the pollen that the male plant produces- this is how they essentially "have sex" and reproduce. The reproductive organs on a female plant will begin growing somewhere around the fourth week of the plant's growth cycle, but for some plants, it may take around six weeks instead.

The Flowering Stage Continued

You do not want to prune your plants any time after this stage of growth, as it will disturb the natural growth of your mature plant and will disrupt the hormones in your plants, which can lead to plant growth complications. You will also need to support the buds of your plant as they can weigh down the stems, and this could eventually lead to the stem snapping. In order to avoid this, you will need to trellis your buds.

This means that you can support your plants using wire or wooden rods that the stems and buds can grow around so that the weight of them is not fully supported by the stems of the plant. At this stage, you also want to continue to feed your plants with nutrients so that they continue to grow strong and healthy, especially in this flowering stage.

Your buds will begin to grow the most rapidly at the end of this phase of plant growth, near the sixth or seventh week. At the beginning of the flowering phase, you will not notice too much bud growth, but as this phase progresses, the growth of your buds will increase exponentially. Then the growth of the buds will slow down again at the end of this phase, which marks the end of the cannabis plant growth cycle.

At this point, the buds will be fully grown and formed to their mature size and shape. When the buds become fully matured (at the end of this phase), your buds will be ready to be harvested and then eventually sold, smoked, or whatever else you plan to do with them.

Testing for Maturity

When it comes time to harvest your plants, there are some different options available to you. In order to determine whether your plant is ready to harvest, you can do so by examining some specific areas of your plant. You must decide whether your plants are ready to harvest so that you can plan for your next stages in the following two ways;

1. Examine the Stigma

This is usually done around the end of the flowering phase when you determine that your buds are mature enough. In order to determine this, you will first look at the stigma of your plants. The stigma are those small female sex organs that are used to attract the pollen released by the male plants. You will need to be examining the stigma on your plant over the course of the flowering phase so that you will be able to determine when they have reached full maturity. The stigma will initially be white in color, but as the growth progresses and the plant matures, you will see them turn to an orange color instead.

2. Examine the Trichomes

The second thing you can look for are the *Trichomes* of your cannabis plant. The trichomes are those small crystal-like structures that are on the buds and the leaves of the marijuana plant.

They are sticky in texture and are actually the part of the plant that gives each strain its characteristic scent that you have likely grown to love. The word "trichome" means "Fine Outgrowth," which is quite telling as to what their purpose is on the plant.

It is in the trichomes that you will be able to find the most information about whether your plant is ready for harvest or not. In order to determine this with the most accuracy, you will need a small, handheld microscope. You can likely buy this from any plant store in your area. This is because you will need to observe the color of the trichomes, which can only be seen with accuracy very close up and cannot be fully examined with the naked eye. You can use this microscope to observe the changes in the color of the trichomes throughout the flowering stage so that when your buds have matured, you can examine the trichomes to double check whether your plant is ready for harvest.

At the beginning, the trichomes of the plant will be clear in color. You will see them transition to an opaque shade, which is an indication that the plant has reached the maximum THC potency levels. At this point, your plants will have reached the maximum potency (in terms of THC content).

As the trichomes break down during the progression of maturity of the plant, this results in a chemical called *Cannabinol* being produced, which has been shown to be one of the reasons why the plant comes with so many pain-relieving benefits, as well as relief from insomnia and inflammation. This chemical does not lead to a feeling of being high; however, that is the THC alone. At this point, you will then see the trichomes transition to an amber shade.

Drying the Buds

Once you have determined that your plants are ready to be harvested, the next step that you will need to take is to dry the buds so that they are ready to be consumed in whatever way you wish to consume them.

The best and most effective way to dry your cannabis buds is thorough a process called *curing*. This process is a process of drying that occurs over an extended period of time. There are a few reasons why curing is the best option for drying your buds. The first reason is that it increases the potency of your cannabis buds. Once you harvest your buds, they do not stop their process of maturity; in fact, they will continue to increase in potency for some time if they are dried through the slower process of curing.

This happens as the non-psychoactive chemicals in the buds are converted to THC over time, which makes the THC content stronger.

If you choose to dry your buds quickly and in higher temperatures, the quality of the smell of your buds will decrease. One of the best qualities of cannabis for many users is the smell of the buds, and many people turn to this quality in order to determine their preference when it comes to choosing a strain. If you cure your buds at lower temperatures over time, you will maintain the strong scent of the cannabis that so many people look for.

Further curing your cannabis buds will help them to last longer in storage without going moldy or stale. This is because they will have fully dried out over a long period of time, allowing you to keep them stored in an air-tight container for up to two years without having them go bad.

You may now be wondering how you can cure your cannabis buds. While there are many methods for doing this, the simplest and most popular method is to cure them at a temperature of 60 to 70 degrees Fahrenheit in a dark room. You will also want to keep the humidity levels at about 50%.

To begin, you will cut off branches of your cannabis plant that are about 12 to 16 inches in length. You will then take off any leaves that you do not need or that are in the way of your buds. Then, you will hang these branches upside down by attaching them to a line of string or wire. This will allow you to leave them to cure over time. You will then set up a fan in the same room, in order to help you circulate the air around the room.

You will leave them in these conditions until you notice the following; Check the leaves of the branches that you have hung up, and when they snap upon being bent (rather than simply folding over), then your plants are almost fully cured. This could take somewhere between 5 and 15 days. When you notice that this point has been reached, you will then be able to remove the buds from the branches.

Once you remove the buds, do not throw away the leaves. The next step is to place the buds in some type of airtight container. In this container with the buds, you will place the leaves. The purpose of this is to allow the moisture content within the leaves to be absorbed by the outer layers of the buds. This will make your buds presentable and aromatic. You will do this for the first 4 to 8 weeks of the life of your cannabis plant.

For the first week, ensure that you are opening the container to allow the buds to get some air at least a couple of times per day. For the second and third week, you can reduce this to once per day. In the third week, your buds are theoretically ready to be used, but leaving them for an additional 4 to 5 weeks will result in a very successful curing process. Whether you have the time for this or not will depend on the purposes for which you are growing your cannabis and whether you have the patience to wait the extra few weeks.

At this point in the book, I would like to ask you to leave a review if you are enjoying this book so far! This will allow others to discover the information contained within these pages and to benefit from it as well!

CHAPTER 5: THE LOGISTICS OF GROWING MARIJUANA

In this chapter, we are going to look more specifically at the logistics involved when you begin growing your own marijuana plant. We will look at how much you can expect this to cost you, what equipment this will require, and other things that you should keep in mind when you begin growing your own plant.

What You Need in Order to Get Started Growing Marijuana in Soil

In this section, we will look at what you will need to grow cannabis in soil. One of the positives of growing your marijuana plant in this way is that you do not require too much equipment, as this is how marijuana was originally found growing without any human intervention. Below are the different items needed to grow marijuana outdoors in soil.

- Fertilizers And Soil

Soil, at its core is a combination of earth, organic materials, as wella ssome clay and some rock. Soil naturally contains elements that contrinute to plant growth, which is why you find plants growing naturally in the wild with no help from humans.

When it comes to choosing your soil mixture, there are a few things that you should keep in mind. For most regular plants, they are found growing in one of three common soil types. These soil types are below.

- Sandy soil
- Slit soil
- Clay soil

When it comes to growing cannabis in particular, it requires a certain soil type to grow and thrive. This soil type is a mixture of the three common soil types seen above. This soil type is called *Loam Soil*. The texture of loam soil is somewhere between sand and clay, so it will compact but not hold its shape too well once compacted.

In order to choose your soil from a store, you can purchase any high-quality soil, but you must make sure that the soil you choose doesn't have any "artificial extended release fertilizer" contained within it (such as Miracle Gro). This type of soil is not ideal for growing high-quality cannabis because it does not contain the proper nutrients for growing a great cannabis plant. If you are new to growing cannabis, the best choice for your soil will be a pre-fertilized organic soil. This is often also called 'super-soil.'

This soil can be used to grow your cannabis plant from beginning to end without needing to add any nutrients, which makes it a great choice for beginners, who will already have a number of things on their mind whn growing their first cannabis plant. You can purchase super soil in a pre-made form from a number of different suppliers, or you can opt to make your own if you are a little more experienced.

If you want to make your own super soil mix, you will need to get your hands on the following ingredients (or some form of them that is accessible to you).

1. A regular potting mix

This will act as the base for your soil

2. Vermicompost

This is a mixture that is made of worm casings that is great for your soil's natural nutrients that will help your plant to grow.

3. Sources of Nitrogen, Magnesium, Potassium and Phosphorus

These elements can come from a variety of different sources, so check your local gardening stores and see what is available in your area.

Once you have added your ingredients together, let it sit for a couple of weeks and it will be ready to help you grow healthy cannabis plants.

- Water

You will need a water source such as a hose or a spray container in order to ensure that your plants are getting adequate hydration.

- Marijuana seeds

You will need to get your hands on some seeds or some cloned marijuana plants in order to begin your plant growth.

What You Need to Get Started Using Hydroponics

Let's take a look at all the equipment and supplies that you will need to run your own hydroponics system. There are numerous different types of hydroponic setups, all with their own unique costs associated. I will walk you through what a standard hydroponic system requires, and I will include the costs associated with this so that you can get an idea of what this will cost you, in order to help you decide what kind of system you will be using to grow your own plants.

- Exhaust Fan

Plants require a lot of fresh air to thrive, and we all learned in grade school, carbon dioxide (CO_2) is required to complete the photosynthesis process. This will require you to have a stable stream of air that can flow throughout your grow room. This can be easily achieved by placing an exhaust fan at the top of your grow-area that will help you remove warm air. Choosing the right size for your exhaust fan greatly depends on how much heat is being created by your lighting design and the size of your grow area.

If you are someone who lives in a region with a warmer climate, you may need to run your lights at night to try to keep the temperatures down. I advise you to set up your lights, turn them on and leave them on for a while. This will help you determine how much airflow is required to keep your grow space at a comfortable temperature for your specific plants. By doing this, you can properly select an exhaust fan that meets your requirements.

- Dehumidifier

As an alternative to an exhaust fan, you could use a dehumidifier. You can create your own artificial sealed environment with the use of a dehumidifier, a supplemental CO_2 system, and an air-conditioner to keep the moisture and the heat levels down.

However, this system is very expensive and is usually not recommended for first time hydroponics growers before they have had at least a few successful yields first.

- Thermostat Controls

After you choose which climate control equipment and lights you want to use, you can start to automate their functions. There are expensive and sophisticated equipment available that you can purchase to control CO_2 levels, humidity, temperature, and lights, but for a beginner, this may not be necessary. All a beginner like you will need is an adjustable thermostat switch and a 24-hour timer for your fan. Using basic thermostat models, you can easily use your thermostat to set the desired temperature for your grow area and then connect your fan into it. As the temperature begins to rise, your fan will turn on automatically until the temperature falls a few degrees under the threshold. This will save electricity, energy, and help you maintain a steady temperature in your grow space. Since you are likely not spending all day in your grow area, using a thermostat with a memory feature will be useful for you when keeping tabs on the environment in your grow space. These devices are small and inexpensive and will show you the humidity level, current temperature, and the readings of the highest/lowest temperatures for the period of time.

- pH Test Kit

It is recommended to own a pH test kit and keep it handy, so you can check the pH of your growing medium, your nutrient solution, and your water. Depending on your plant, it will thrive in certain pH ranges, and you want to ensure that you are maintaining this. If your pH rises or falls out of the range, you can cause something that is called 'nutrient lockout.' This means that your plants won't be able to absorb the nutrients that you are giving it, so making sure that your water and soil are at the right pH levels consistently is important. Make sure you are also testing the nutrient mix that you are giving to your plant to make sure it is within this desired range.

- Growing container

Choosing the container that you will grow your plant in will be highly dependent on the size of your plant, the hydroponic system that you have chosen to use, and the growing medium that you will be using.

For instance, a flood-and-drain, tray-style hydroponic system may utilize small net pots that are filled with clay pebbles or even just a big slab of rockwool to grow several little plants. On the other hand, a 'super-soil' grow may use a variety of 10-gallon nursery pots to grow a few large-sized plants.

If you are looking for the least expensive option, you can choose between cloth bags or perforated disposable plastic bags. Some people may decide to purchase 'smart pots' and spend a little more money on them. Smart pots are containers designed to better the airflow for your plant roots. The most common medium for first-time growers is the simple five-gallon bucket. The key here is drainage. Since plants are sensitive to conditions that contain too much water, as they can become water-logged, you must be sure to drill holes at the bottom of the bucket and set the bucket on a tray in order to allow excess water to leave the medium to avoid this problem.

- Nutrients

Your plant will require you to add nutrients to it as there is no soil involved. Your plant will require the following macronutrients: Nitrogen (N), Phosphorus (P), and Potassium (K). You will also need these following micronutrients but in smaller quantities: Copper, Calcium, Iron, and Magnesium. Usually, macronutrients sold in stores are sold in two-parts to help the elements from precipitating out of the solution, as this would cause waste. This means that you will need to purchase two bottles of nutrients for the plant's vegetative state and two bottles of nutrients for its growth stage as well as a bottle of micronutrients.

Besides that, there is a possibility that you may need to purchase a Cal/Mag supplement because some specific plants will need more magnesium and calcium when compared to other plants. This will depend on the specific plant that you are growing.

When you have chosen your nutrient products, all you have to do is just combine them with water as per the label instructions. Then, water your plants with that nutrient solution in order to feed them. Always start watering your plants with a solution that is half-strength, because some varieties of plants are easily burned if they receive too many nutrients too quickly. In most cases, overfeeding your plants is worse than underfeeding them. Once you have gained more experience with feeding your plants, you will slowly learn how to 'read' your plants for signs of excess nutrients or nutrient deficiencies.

The Cost of Growing Marijuana With Hydroponics

The total cost estimation for running your own hydroponics system for your first grow is approximately: $730. You can see the breakdown of this cost below. Bear in mind that you can opt for less expensive equipment like choosing cheaper lights like CFL as compared to LED.

You can also choose to buy a prepackaged hydroponics starter grow kit that may be slightly easier and cheaper for you to use that costs about $500. It typically comes with a 2' x 2' grow tent, nutrients, digital timer, pH drops, adjustable hangers, lighting, CO_2 bags, and a clip fan. Keep in mind that electricity also costs money, depending on where you live. Although this cost isn't very high, you will notice an increase in electricity consumption when you are starting to grow plants indoors. There are also more sophisticated grow kits out there on the market starting from $1000 that you can purchase. Decide what is best for you and go with it! Make sure you are doing your research and look out for sales and promotions when doing this.

Hydroponics Equipment List

- 3- or 5-gallon bucket, you will need one for each plant you grow ($20+)
- Clay pellets (enough to fill the bucket) ($30+)
- Grow table ($75 - $150+)
- Rockwool cubes (one 1.5inch starter plug per plant) ($10+)
- Water pump (as big as possible) ($20 - $50+)
- Reservoir tank (depends on the size of your garden) ($25+)
- Air stone ($6+)

- Air pump ($12+)
- Plastic tubing ($5+)
- Drip line ($15+)
- Drip line emitters (1 – 2 per plant) ($12+)

Additional Equipment

- Seeds of your choice (depending on what you wish to grow) ($10+)
- Grow tent (2' x 2' or 3' x 3') ($120+)
- Nutrients ($50+)
- Lighting equipment of your choice:
    - HID ($200+)
    - LED ($200+)
    - CFL ($50+)
- Carbon filter ($100+)

CHAPTER 6: MARIJUANA PLANT MAINTENANCE

In this chapter, we are going to look at what you need to do in order to maintain your marijuana plant. Properly maintaining your garden will help you to get the highest yield possible and keep your plants healthy and thriving in their growing environment. Depending on what type of growing method you have chosen, there are slightly different ways to maintain your plants. However, there are a general set of rules you should always be following in order to achieve the right level of pH, the right temperature levels, and to prevent things like pests and disease. I will be walking you through all the necessary maintenance that needs to be done for your garden so that you can prevent anything negative from happening to your plants.

Maintenance That Needs to Be Done

There are several areas of maintenance that you constantly need to be looking out for in your garden. None of these tasks are exciting in the least, which means that they generally are given the least attention. Everybody loves to maintain the plants, grow beds, and their garden, but the actual maintenance of your reservoir is typically left as an afterthought.

Although a little bit boring, your garden maintenance plays a big role in the success of your system, so not properly maintaining it can cause death to all your plants in a worst-case scenario situation. Let's take a look at the six areas you need to maintain on a frequent basis.

Temperature

The temperature is important for both soil-based and hydroponic systems. When growing both indoors and outdoors, this is important too. When growing your plants outdoors, the climate will determine the temperature of your plant growing environment, so you must ensure that the time of year and the climate in which you live are not going to cause problems for your plant at any stage of growth.

In a hydroponic system, you ALWAYS need to keep your water/nutrient solution at a temperature of 65 to 75 degrees Fahrenheit. If temperatures exceed 75 degrees, the levels of oxygen are at risk of decreasing, which will create the ideal environment for root rot to grow. On the other hand, low temperatures will cause the plant to grow slowly. If you have a small reservoir, you can use aquarium heaters to warm up your nutrient solution. You may require a stronger heater if your reservoir is quite large. When it comes to your solution becoming too warm, you can cool your nutrient solution in these following ways:

- Use a reservoir chiller

The easiest solution to cool down a warm nutrient solution is to use a water chiller. These chillers will help you run your nutrient solution through a coil that is cold, which will cool down the temperature of it. There are some DIY options for this, but water chillers are much better as they are accurate and lets you choose your ideal temperature for your solution to be at and can help you keep at a constant temperature. This piece of equipment varies in price depending on how big your reservoir is as the bigger it is; you require a more powerful chiller.

- Keep your reservoir in the shade

A cheap way for you to keep your nutrient solution is cool is to build some sort of cover to help keep the temperature cool. You can use a shade cloth or build a box that can cover your solution. This will also help keep the sun away as some plants thrive well with too much direct sunlight.

- Add ice to your reservoir

A cost-effective alternative to cooling down your reservoir is to add ice into your nutrient solution. The easiest way to do this is to freeze your own ice by filling up a jar with water and freezing it, then you can add it to your reservoir.

However, you have to keep in mind that you don't want to cool your solution too quickly or by too much as that can harm your plants. Make ice blocks/cubes of different sizes so you can control the amount of extra water and temperature change that you're giving to your solution.

- Change the color of your reservoir

If your reservoir is a darker color, it is likely to hold more heat compared to a reservoir that is of a lighter color. A good way to get rid of some heat is to paint your reservoir into a lighter color like white. This can help you lower the temperature of your solution by a few degrees. If you don't want to paint your reservoir, you can simply wrap it in mylar or foil as metallic will help deflect some heat away.

- Top your reservoir off with cold water

If your solution is increasing in temperature on a warm day, simply add some cool water into your solution. This is likely the easiest way to lower the temperature quickly. Do keep in mind if you are using this method to make sure to top up your solution as it will likely dilute due to the extra water being added in.

- Bury your reservoir underground

If your hydroponic system is located outside, you can simply bury your reservoir underground with some soil. The natural ground will keep your reservoir cool by keeping it away from direct sunlight. This is also a very cheap option but will require more manual labor.

Oxygen

To achieve healthy root growth in your plants, you need to have a nutrient solution that is well-oxygenated. This is an absolute must for your plant to grow and helps with the growth of beneficial organisms that will strengthen your overall plant. You can also add an extra air stone, which will help your solution to maintain a higher level of dissolved oxygen.

Water Levels

If you are using a circulating hydroponic system, you need to make sure you are topping off your reservoir when you are changing the solution. A lot of water will be lost due to plant processes and evaporation, so you will need to keep replacing the water to keep it running smoothly. The smaller your reservoir is, the more often it will need to be topped up. Depending on which hydroponic system you chose to use, you will need to change out your solution based on that.

There are a lot of different and conflicting opinions out in the market, so it will be hard for you to find one answer for this. In my opinion, I find that changing your nutrient solution every two weeks is ideal for your plants. If you want to be more accurate, you can purchase an EC meter that will help give you an analysis of how much fertilizer is currently in your water. However, it won't tell you the amounts of each nutrient in there. Plants do not absorb every nutrient at the same rate, so topping up too much of one nutrient can also harm it. Water refreshes will allow you to ensure that your solution is well-balanced in the nutrient that it contains. This will help you achieve less build up in your system, a good chemical balance, and will allow you to clean your entire reservoir.

Adding a filter into your hydroponic system is always a good idea. What this does is it will prevent things like debris and plant matter from wandering into your reservoir. If you clean the filter often, you can reduce the amount of build-up which may attract pests. However, if you are using a DWC system (deep water culture), you don't need to use a filter as debris will typically float to the top of the solution. However, you do need to clean that debris out.

Lighting Cycles

Lighting is very important, and it can have a large effect on your plants in the long run. The timing that you set for your light/dark cycle is extremely important when you are growing a plant indoors, as there is not a natural cycle like there would be outside with the sun. Ideally, your lights need to be on for 16 – 20 hours over a 24-hour period during the vegetative growth stage. Then, you will need to switch over to 12 hours of light per 24 hours during the time that you want them to bloom. You will need to turn your lights on and off at the exact same time every day; otherwise, you may stress out your plants. Having a timer is essential for indoor growth. You can also use your timer for your exhaust fan as well, but it is easier to just spend a couple of extra dollars on a thermostat switch.

Tools and Cultivation

Regularly checking the functionality of your equipment is always a good idea. Things like broken aerators, pumps, and connections are things that typically go unnoticed, which can cause your plants harm or prevent them from growing. If you can, keep spare parts around for backup. This will help prevent any long-lasting damage if a certain part breaks down at an inopportune time.

CHAPTER 7: HOW TO AVOID PESTS AND DISEASE

The best way to prevent pests and diseases in any of your plants is to get familiar with the most common ones. By understanding what they are and the environments that they typically like, you can prevent it from happening rather than trying to fix it every single time. Typically, it should be very easy to identify what is happening to your plant as long as you have existing knowledge of it. If you don't, the signs and symptoms are likely to go unnoticed until your plant starts to die.

Common Marijuana Plant Pests

Let's first take a look at the most common pests that you could encounter in your garden.

1. Spider Mites

Spider mites are the most common and annoying of all indoor plant pests. They are small bugs that are less than 1mm in length. Technically, they are in the arachnid family, and because of their small size, many people don't notice that they have them until your plants are very damaged. In order to fix this problem, you need to be able to spot them before they do too much damage.

You can do this by looking for webbing; if there is webbing, you may have a case of spider mites. Another way to identify this is to grab a tissue and wipe the underside of your plant leaves. If you notice your tissue is streaked with spider mite blood, then you 100% have the spider mites.

2. Thrips

Thrips are tiny bugs that are around 5mm in length. They are a bit harder to identify, but if they are there, their damage is quite obvious. Look at the tops of your plant leaves and see if you can see any small metallic black specks. If there are, you may notice that your plant leaves are starting to dry out and turn brown. There also may be some yellowish-brown spots too. This is happening because the thrips are sucking the leaves dry of their moisture.

3. Aphids

Aphids are also commonly known as plant lice. They can be gray, green, or black. It does not matter what color the aphids are; they all do the same damage to plants. They typically like to suck all the moisture and juice out of plant leaves, which turns them yellow. They typically will gather at the stem of your plant, but you can find them anywhere. Check out the stem first if you suspect that you have an aphid infestation.

4. Whiteflies

Whiteflies are flies that look like white moths and are about 1mm in length. Since they are white, it makes them pretty easy to spot, but they are difficult to get rid of as they will fly away pretty quickly if the plant is disturbed. They also like to suck all the moisture off of plants, which will cause your plants to have some yellowing and white spots.

5. Fungus gnats

Did you know that adult fungus gnats aren't actually harmful? However, their larvae will eat your feeder roots and roots, which will slow down your plant's growth. This will also cause bacterial infection and could lead to plant death.

Remember that the best way to fix your pest problems is to prevent them in the first place. I will teach you prevention measures in the next subchapter.

Common Marijuana Plant Diseases

Now, let's take a look at some of the most common indoor plant diseases. Understanding these will help you quickly identify them when it happens so you can help your plants recover as fast as possible.

1. Powdery Mildew

If your plants look like somebody had sprinkled white powder all over your stems and leaves, then you probably have powdery mildew. If you leave this untreated, it will stunt your plant growth, causing yellowing of plant tissues and leaf drops. If let for a long time untreated, your plant will die.

2. Downy Mildew

Downy mildew and powdery mildew are different, so don't get these two confused. While powdery mildew causes your plants to have white powder all over your stems and leaves, downy mildew will mostly show up on the underside of your leaves. This doesn't look as powdery as what powdery mildew looks like, but both of these mildews will cause your plants leaves to turn yellow. Just remember, if the white powder appears on the underside, it is likely downy mildew, whereas if the powder appears on the stems, it's likely powdery mildew.

3. Gray Mold

Gray mold is also known as ghost spot or ash mold. To identify this disease, look for fuzzy gray abrasions that typically start out as little gray spots on your leaves. You will notice that these spots continue to deteriorate until your plant turns mushy and brown.

4. Root Rot

If your plants are being given too much water with pathogens in it, root rot will happen. We talked about this earlier in the book, which is why maintaining proper levels in your hydroponic medium is crucial. When this happens, your plants will begin to wilt and turn yellow. You will notice that the roots will get mushy as well.

5. Iron Deficiency

If your plants lack iron, it means that they also lack chlorophyll. This is shown when leaves turn into a bright yellow color while the veins remain green. Often times, this is improperly diagnosed as another disease when simply your plant is just lacking some iron.

How to Avoid and Prevent Pests

You likely know by now that gardening comes with its challenges, as there are many factors involved. One of these factors includes pests and disease, which could plague your plant when you least expect it. In this section, we are going to look at how you can prevent pests and disease in your hydroponic garden so that you end up with the biggest and healthiest yield possible. Now that you know the most common pests that your plants could be facing, we are going to look at what you can do to prevent this.

As I briefly mentioned, the number one factor that will help to keep your garden healthy and pest-free is preventing the chances of pests wanting to inhabit your garden in the firsts place. By taking the appropriate steps to make your garden an environment that pests want to stay away from, you will give yourself the best chance for a healthy and thriving hydroponic garden. We will look at what these preventative measures are below. We will begin by talking about pests. Specific to the pests that I mentioned earlier in this chapter, there are a few things you can do to help control and prevent their appearance in the first place.

- Sticky Traps

Firstly, you can use sticky traps. You have probably seen these traps in your lifetime before, as they are commonly used to trap house flies. You can also choose different colored traps to help you better identify which pests you are catching. For instance, yellow card sticky traps will attract whiteflies and fungus gnats, while blue card sticky traps attract thrips. When using this method, be sure to place some of your traps at a low level and some others at medium level in terms of your plant height. This is where the fungus gnats like to congregate, so this will give you the best chance of catching them.

- Pest Sprays or Pesticides

You can also use different types of sprays to get rid of these pests, but be sure to avoid chemicals like *Eagle* or *Avid*. This is especially important if you are growing plants that you later plan to eat or ingest in some form. Try to choose organic sprays if possible. Alternatively, many people like to make their own organic pesticides. These are especially beneficial if you are growing vegetables or if you are against the use of pesticides. Some common homemade pest sprays include the following;

To deal with spider mites, many people choose to make a spray consisting of Himalayan pink salt and water, which they then spray on the plants.

Another popular natural homemade pesticide is made from 1 ounce of garlic and 1 medium onion mixed with 1 quart of water, which are left for one hour to infuse. Then, 1 tsp of cayenne pepper and 1 tbsp of liquid Castile soap (or vegetable soap). This can then be sprayed on your plants to get rid of a wide variety of pests without harming the plants or vegetables themselves. Arguably the most well-known natural pesticide is called *Neem*. This has been used for years by aboriginals to keep pests of all sorts away.

This pesticide gets its name from the leaf of a tree, and you will need to get your hands on oil from this tree leaf in order to make it, though it should not be hard to find due to its popularity. This pesticide is very powerful while also being completely natural, which makes it the top choice for many. To make this, you will need to mix one-half ounce of Neem oil with one-half teaspoon of organic soap liquid as well as two quarts of water (warm). You can then spray this on your plants without harming them, and this will keep pests away.

- Beneficial Predators

Lastly, you can use beneficial predators to kill pests that appear on your plants. One example of these beneficial predators are *Nematodes*. Nematodes are more commonly known as roundworms. These beneficial predators can be beneficial in treating pests such as gnats, caterpillars, grubs, crane flies, thrips, ants, moths, and fly larvae. One of the benefits of using these beneficial predators to treat your plants is that they will not harm you or your plants, and there is no risk of putting too many of them in your garden. This means that they will be able to eradicate a pest problem or prevent it in the first place, without risking your plants themselves.

How to Deal With Disease

Once again, the number one factor that will help to keep your garden healthy is preventing the chances of diseases wanting to grow or inhabit your garden in the firsts place. By making your garden environment uninhabitable to diseases like mold or fungus, you will keep your plants healthy.

A good idea for your grow room is to always ensure a constant, light breeze is present, as this will strengthen the stems of your plants and will help you to create an environment that is less hospitable for flying pests and mold. You can use a wall-mounted circulating fan for this. Make sure to prevent windburn of your plants by not pointing your fans directly at your plants.

Before you begin growing anything or introducing plants into your garden, ensure that the entire apparatus and other components of your garden are sterilized and disinfected. You can do this by using a ten percent peroxide solution to disinfect everything before you begin gardening. Preventing pests and diseases is much better than trying to fix your plants after the fact. First of all, wear clean clothes before going into your grow room. Many diseases and pests could be on your clothes from the outside world and can easily make a new home in your plants.

Ensure that you are also wearing indoor shoes that haven't touched the outside world to prevent bringing in any outdoor pests or diseases. Next, make sure you are keeping your grow area clean by cleaning up any spills or accidents. Many mildews and molds are caused due to excess humidity and water, so ensuring that you have clean water is the best way to prevent these nasty molds from growing. Lastly, keep your plants as clean as possible. If you come across any dead plant matter near your grow area, pick it up. Be sure that you are also pruning your plants whenever it is necessary so you can remove dead/diseased leaves and branches. By reducing the amount of dead plant matter, you are reducing the risk of getting any pests or diseases.

In addition, keeping your plants as healthy as possible is another great way to fight off diseases and pests. This is why choosing the nutrients you use for your plant is important. For instance, make sure you are always keeping tabs on the pH levels of your plant, the more optimal your pH levels are, the better your plants can fight off any diseases and pests. Be sure to also do some research on the best nutrients for the type of medium you are planning to use. If you plan on using coco coir as your hydroponic medium, then look to purchase coco-coir specific nutrients.

Since coco coir binds with magnesium and iron, your plant can easily become starved of those nutrients because the base won't contain the optimal amounts. Moreover, giving your plants supplements will also keep them healthy. Rhino Skin is a supplement that is a soluble potassium silicate formulation that will help strengthen your plants in a physical way. Essentially, it strengthens your plant's cell walls and helps protect your plants from disease, drought, heat, and other possible stressors.

CHAPTER 8: HOW TO GROW HEALTHY CANNABIS PLANTS

Common Mistakes to Avoid When Growing Marijuana

When you begin anything new, you can make mistakes, and they will teach you as you go, but when it comes to hydroponics, it can be devastating to make a mistake that can cost you your entire crop. In this section, we will discuss several of the most common mistakes made by beginners so that you can learn from the mistakes of others instead of having to go through them yourself.

Forgetting pH

When growing crops in a hydroponic system, the pH of your system is very important as your plants will need to be provided with very specific pH ranges in order to grow and remain healthy and alive. The nutrient solution that you give your plants needs to be at the optimal pH for those specific plants so that they are not exposed to an environment that is too acidic or too basic for them. To avoid this mistake, ensure that you are aware of the pH level that your plants need to remain at.

Then, you will need to have some way to measure the pH of your solution. You can do this with a pH meter or with pH testing strips. You will need to test the pH of your solution once per day at minimum, if not more. If you are new to hydroponic gardening, you may wish to check this more often to ensure that you are on the right track until you get more comfortable with it.

Lighting Errors

As you know, by this point, lighting is one of the most important factors present in your hydroponic garden, next to the pH level. If you are going to choose one place to invest the most money, choose good quality lighting. Without the correct lighting schedules or the right types of lightbulbs, your plants could have a difficult time growing. Ensure that you are aware of the specific lighting schedule and type of lighting that your plants need and then make sure that you are giving them exactly what they need in terms of light quality and duration. Keep in mind that light streaming through a window is not enough to provide your plants with the light they need to grow strong and healthy.

Overfeeding Your Plants

In most cases, overfeeding your plants is worse than underfeeding them. Once you have gained more experience with feeding your plants, you will slowly learn how to 'read' your plants for signs of excess nutrients or nutrient deficiencies.

Improper Nutrients

One major mistake that beginners often make is turning to a traditional fertilizer that can be found at your nearest nursery. The problem here is that these are often optimized for soil-based gardens and not hydroponic systems. In order to ensure that your plants are actually benefitting from the nutrients that you are giving them, you must ensure that these nutrients are fully dissolving in the water so that they are creating a nutrient-rich solution. If they simply sink to the bottom and refuse to dissolve, they will not be passed onto your plants through the roots, and your plants will then likely suffer from a lack of nutrition.

Tips and Tricks for First Time Hydroponics Growers

If this is the first time you are growing plants indoors and in a hydroponic system, the first thing you have to do is set up the right space to do it.

This space doesn't need to be a typical looking grow 'room' like the ones you see in movies. It can simply be a spare room, a cabinet, tent, closet, or the corner of your basement. The most important part is to tailor your equipment and the plants you choose to fit the space. This is why choosing the right indoor plant is important as some plants grow to be too large for a small tent.

For your first project, you ideally want to start with a smaller plant in a smaller place. The smaller your grow, the less expensive it will be for you to complete your project. It is also way simpler to watch over just a few plants rather than a large amount of them. In addition, smaller grows, in the beginning, will be less costly for you if you do end up making mistakes. Keep in mind that a lot of novice plant growers will experience obstacles like losing a plant to disease or pests. If you are trying to grow ten plants at once in your first grow, you may end up losing a lot of money if they are unsuccessful, whereas if you started off with just 1 – 2 plants, it won't hurt your wallet as much. Although we are trying to keep your growth area small, we must also think big at the same time. When you are designing and creating your grow space, make sure to account for the room that the equipment will need like fans, ducting, and lights, as well as the room that the plants will take up.

If your grow room is somewhere quite small like a closet, tent, or cabinet, you can choose to open the door or the entrance and take out your plants when you are working with them for ease of access. If this is not possible, make sure you are accounting for the room that you will need to take up in order to physically care for your plants.

Next, make sure you are maintaining cleanliness. This is especially important when you are growing your plant indoors. Make sure you are frequently sanitizing your workspace and avoid growing your plants on raw wood, drapes or carpeting as these types of spaces are difficult to clean. Another important part of your grow space is to ensure that your light is sealed tight. If you leak light during dark periods, it can confuse your plant as it will complicate the imitation of the light and dark cycles of the sun that you are trying to emulate. This is why it is recommended to put it in a sealed closet or a tent that can be zipped up to keep the growing environment dark when need be.

Further, you should ensure that your grow space is placed in a convenient spot as you will need to monitor your plants often and carefully. It is important to check in on your plant every day, and if you are a beginner, you probably would like to check in multiple times in a day until you're sure everything that is needed to be done is completed.

If your grow space is in an inconvenient area like in a separate building, it may prevent you from being able to check in on your plant multiple times per day. Additionally, you have to maintain optimal temperature and humidity in your grow space. If your space is humid or warm, you may be faced with more obstacles related to grow-space control. Choose a dry and cool area with plenty of fresh air that is highly recommended.

CHAPTER 9: WHAT TO DO ONCE YOUR PLANTS HAVE GROWN

Cloning is the final step of your marijuana plant growth. Once your plants have grown and you have determined that they have reached full maturity, it is time for you to decide what you are going to do with them next. One of the common next steps is cloning, which we will discuss in this chapter.

Cloning

Cloning is also known as *Asexual Reproduction* since it involves only one plant instead of the reproduction that we discussed previously, which combines the genetic material of two plants through pollination. In order to clone a plant, you will use a razor which you will use to cut off a piece of your original plant. The plant which you wish to clone is called the *Mother Plant*. This plant is chosen for its health, the product it produces, and the ease with which it was able to grow since these characteristics will be passed onto the new plant in which you will grow through cloning. When cloning a plant, you want it to be somewhere around eight weeks into the vegetative stage of plant growth. This is the opportune time for cloning.

Before you are going to clone your plant, avoid giving it any fertilizer for a few days before cloning in order to prepare it. You will want to choose a branch that is lower on the stem, and then you will use your razor to cut off the branch at a forty-five-degree angle, as this gives you enough surface area for the new plant to grow new roots from. Once you make the cut (do not use scissors as they are not sharp enough and will crush the stem), put the branch directly into water. This will ensure that it stays fresh and that no air bubbles will be present inside of the branch.

Once you do this, you will then clip off the leaves in order to help induce growth in this new small plant. Halfway down the stem of the leaves, clip them off. After you have done this, you can plant your new clone. You can plant it in soil that has no nutrients (it doesn't need them yet), in water alone or in Rockwell. Any of these three mediums will do.

Once you start to see roots growing on the bottom of your clone's stem, you will be able to plant them in whatever location you plan to have them grow from here forward. Then, you can continue the process of growth as you normally would from here on. This is why it is called a plant life cycle; it goes around and around.

Once you have grown your first set of cannabis plants, you can determine which ones you want to clone so that your next set of plants will be as strong as possible. By choosing which plants you want to clone, you can ensure that your cannabis plant production improves with every growth of new plants. For this reason, it is important to pay close attention to which of your plants sprouted the quickest, formed a seedling the quickest, and flowered the quickest. This is an indication of a strong plant and a strong seed.

When it comes to commercial cannabis production, the farmers will usually plant all of the seeds and then choose the best plant that results. When that ideal plant creates its own seeds, the farmer will then clone those seeds to create a harvest comprised fully of ideal plants.

CONCLUSION

Over the past few decades, marijuana has grown in popularity and has even become decriminalized and legalized in numerous countries. With the marijuana liberalization movement, many people have become curious about this plant that offers so many healing properties.

As you can now understand, the history of marijuana is wrought with controversy. To this day, there are still many conflicting opinions on this plant. I hope that this book provided you with a deeper understanding of the many sides of the marijuana plant, including the medicinal properties it boasts, the ways that it can be used and benefitted from recreationally, and what the future of marijuana may hold. I also hope that this book opened your mind and helped you to discover for yourself how you can benefit from this wonderful plant that is full of possibilities for anyone, no matter what they are looking to get out of it.

The purpose of this book was to help you get an understanding of the basics of marijuana in order to help you learn to grow your own marijuana plant in the most effective and successful way possible.

This book guided you through the history of marijuana, the science behind it, its effects, ways to use it, different strains of marijuana, its plant lifecycle, and things that you should know before you grow your own plant to ensure its success.

Not only are there numerous uses for the marijuana plants that you can grow using the information you have learned in this book (ranging from food to various therapeutic uses and properties), but it is also extremely fulfilling to raise your own marijuana plants from seedling to harvest. You will also be able to put your knowledge to the test through trying what you have learned in this book on your own, and you can use this information to grow a wide variety of different marijuana strains of your choice, depending on your intended uses and your preferences. The best way to improve your skills is to practice and to learn from your mistakes. I wish you success as you take on this journey, and I hope that you continue in your pursuit of the healthiest and most potent marijuana plants as you can grow.

Finally, if you would like to learn more about hydroponics including what can be used for and what types of vegetables you can grow with hydroponics, please check out my other book, *Hydroponics: The Complete Step-By-Step Guide for Beginners to Build a Hydroponic System and Grow Vegetables, Herbs, and Fruit in an Organic Way.*

Hydroponics

The Complete Step-By-Step Guide for Beginners to Build a Hydroponic System and Grow Vegetables, Herbs, and Fruits in an Organic Way.

Garrick S. Thatcher

© Copyright 2020 by Garrick S. Thatcher.

All right reserved.

The work contained herein has been produced with the intent to provide relevant knowledge and information on the topic on the topic described in the title for entertainment purposes only. While the author has gone to every extent to furnish up to date and true information, no claims can be made as to its accuracy or validity as the author has made no claims to be an expert on this topic. Notwithstanding, the reader is asked to do their own research and consult any subject matter experts they deem necessary to ensure the quality and accuracy of the material presented herein.

This statement is legally binding as deemed by the Committee of Publishers Association and the American Bar Association for the territory of the United States. Other jurisdictions may apply their own legal statutes. Any reproduction, transmission or copying of this material contained in this work without the express written consent of the copyright holder shall be deemed as a copyright violation as per the current legislation in force on the date of publishing and subsequent time thereafter. All additional works derived from this material may be claimed by the holder of this copyright.

The data, depictions, events, descriptions and all other information forthwith are considered to be true, fair, and accurate unless the work is expressly described as a work of fiction.

Regardless of the nature of this work, the Publisher is exempt from any responsibility of actions taken by the reader in conjunction with this work. The Publisher acknowledges that the reader acts of their own accord and releases the author and Publisher of any responsibility for the observance of tips, advice, counsel, strategies and techniques that may be offered in this volume.

INTRODUCTION

Congratulations on purchasing Hydroponics: The Complete Step-By-Step Guide for Beginners to Build a Hydroponic System and Grow Vegetables, Herbs, and Fruits in an Organic Way by Garrick S. Thatcher and thank you for choosing this book to begin your foray into the world of hydroponics gardening.

The following chapters will discuss everything to do with hydroponics and how easy it will be for you to begin growing your fresh produce, flowers, and herbs without using one grain of soil. Growing plants in nothing but water is not a new concept, but it has gained a renewed interest in the past few years. Hydroponics gardening might be the answer to the food shortage that is plaguing the world today.

But even without solving the issue of world hunger, hydroponics is a great way to grow fresh produce and herbs for your use at home, no matter how much space you may or may not have. It is relatively easy to do and a fun way to spend time at home. And having fresh herbs for your recipes or fresh leafy greens for your salads is reason enough to learn how to grow things hydroponically. Just as with any other form of gardening, it takes a bit of work to grow things hydroponically, but it is entirely worth it the effort you expend.

There are plenty of books on this subject on the market, thanks again for choosing this one! We hope that you enjoy this book and find that it is packed with plenty of useful information!

CHAPTER 1: WELCOME TO HYDROPONICS

What Is Hydroponics?

By simple definition, hydroponics means working water, so hydroponics gardening means to grow plants in water, without using soil. People have become so used to the idea that plants need soil to grow that growing plants in water seems like an utterly extraordinary concept. But growing plants in water is not unique, and it is not complicated. Plants will not only grow without soil, and they will thrive. Plants often grow faster and better with their roots sitting in water or hanging in very moist air than they do with their roots sitting in hard dirt. Many of the foods that you eat are already being grown hydroponically, so there is no reason why you can't learn the methods employed in hydroponic gardening and use them at home for your own gain.

The process that plants use for growing is called photosynthesis. In this process, the plant uses the energy that it gets from the light to convert water and carbon dioxide into the simple sugar that is known as glucose. The process that makes up photosynthesis uses carbon dioxide, light, and water – but not soil. There is no mention of the word 'soil' in the chemical equation that tells how photosynthesis happens.

Some people say this is definite proof that plants can grow without the use of soil. Plants need nutrients and water, which are both easily obtained from the soil that the plant grows in. But if a plant can get the nutrients and water that it needs from another source, then it will not require soil to grow in. And that is the basic idea behind hydroponics gardening, that plants can grow in water and air, without the use of soil.

During the last century, horticulturists and scientists have been experimenting with different ways to use hydroponics for gardening. One of the possible applications of hydroponics gardening is being able to grow fresh fruit and produce in areas of the world that have little to no soil in which to grow food. One remarkable example of growing food in water is the rice crop. It wasn't until seasonal flooding of the rice fields proved that rice grew better in water than soil that rice became a hydroponic crop.

While hydroponics gardening has been around since the time of the ancient Greeks, an invention in the past one hundred years made hydroponic gardening more effortless than ever before. That invention was plastic. Originally the beds that built the nutrient reservoir for hydroponic gardening were constructed of metal, which rusted, or concrete, which was very heavy and broke easily. With the advent of plastic, hydroponics enjoyed a rebirth.

Plastic water reservoirs and plant pots were so much lighter than concrete, and so much easier to move around when needed. Pipes for transporting water and nutrients could now be made of plastic, and not the lead that used to rust and break.

How Hydroponics Works

In gardening hydroponically, the soil is not used to grow plants in. In this form of cultivation, the roots of the plants are supported in some sort of inert medium. Some of the most common substances that gardeners use for the inert medium are rockwool, peat moss, perlite, vermiculite, and clay pellets. Using a growing medium allows the plant to have access to nutrient-rich water and oxygen at the same time. This allows the plants to grow and flourish, and it is the basic idea behind hydroponic gardening. You can grow a plant without using soil, but all plants have the same needs for the things that the earth provides them. Plants need appropriate nutrients, protection from adverse temperature changes, moisture, and oxygen if they are to grow properly. Many people have grown something hydroponically without even realizing that they are doing it. If you have ever suspended an avocado pit in a jar of water, grown a piece of potato in water, or set the cutting from a flowering plant in water until it rooted, then you have grown a plant hydroponically.

The jar supports the plant, and being indoors protects the plant, while the water provides the nutrients for the plant. One of the most significant issues when growing a plant in water is to get it the amount of oxygen that it needs. And whether a plant is sitting in soil or water, all plants have the same basic environmental needs.

Plants Need Air. All plants need fresh air surrounding them so that they can pull the carbon dioxide out of the air and use it to make energy through the process of photosynthesis. When the plants do this, they return oxygen to the air, since it is a byproduct of photosynthesis that the plant can't use.

Plants Need Moisture. Water is one of the most critical things that every living form needs to be able to survive, and plants are no different. The water in the plant is what keeps the cells sturdy and keeps the plant upright and firm. The water in the plant also carries nutrients throughout the plant, much like blood does in the human body.

Plants Need Light. A good source of light is critical to the growth of any plant, because the plant will need to use the light to draw in energy when it is completing the process of photosynthesis.

Plants Need Proper Temperatures. The temperature level of the air is determined by the heat rising up from the earth and mixing with the air. Plants need a constant air temperature to be able to grow to their fullest potential, and this is something that is not guaranteed when a plant is grown outdoors in the soil. And local temperatures will often determine what types of plants can be grown in that area. When plants are grown indoors, hydroponically, you will be able to control the air temperature to ensure the maximum growth of whatever you choose to grow.

Plants Need Nutrients. The average plant will require a mixture of seventeen different nutrients in order to grow tall and strong. These nutrients will help them to sustain their life cycle and their growth patterns. Plants will find the carbon, oxygen, and hydrogen in the air that surrounds them, but the rest of the nutrients that they need will need to come from the environment that the plant is growing in, mainly the growing medium and the water that moistens the plant. Without the proper mix of nutrients, the plant will not survive.

When a plant is grown in the soil, it will get all of its minerals and vitamins from the soil, so a plant that is grown in water will need to be fed nutrients.

And that is one of the best things about hydroponic gardening, the idea that you will be able to feed your plants the exact mix of nutrients that they need to flourish.

Hydroponic gardening has been in use for centuries and until recently did not receive the level of respect that it truly deserves. While there might be some disadvantages to hydroponic gardening, the advantages of this method of gardening will ultimately outweigh any small problems that you might encounter. And in the meantime, you will be growing fresh herbs, leafy greens, fruits, and beautiful flowers. This method of growing has unlimited potential.

CHAPTER 2: WATER OR SOIL?

If you are thinking about starting a garden at home, you are probably wondering which medium will be the best for you to work with, water or soil. You have the ability to choose whether you want your plants to grow in soil to get their nutrients and then watering them, or if you want to grow your plants directly in the water that is full of nutrients. Deciding whether to grow your garden in soil or water will depend mostly on personal preferences and available space, but growing your plants in water may be better for their overall growth.

Plants need water, nutrients, and light to survive. The water transports the nutrients from the roots to the leaves. Whether the plant is grown in soil or water, without water, the only part of the plant that would be able to take advantage of the nutrients is the roots. Different plants will need different amounts of water. And the weather and the season will also play a role in how much water the plant needs. So while the plant may be able to grow very well without soil, no plant will be able to grow without water. And if you want to grow your plants in soil, they will still need water, and they will still need added nutrients.

No soil is so wonderful that it has the perfectly correct mixture of nutrients for every possible plant life. And this is why growing plants in water is the best choice. There are many other reasons why hydroponic gardening is the best possible choice for any home garden.

Advantages of Hyrdoponic Gardening

Hydroponic gardening is simple yet sustainable. This type of gardening is not simple in the idea that you can just set it up and forget about it. Like soil gardening, hydroponic gardening will require some amount of regular attention. But you will not need a lot of money to start a system, and you will not need to know anything about the local climate or the changing of the seasons. Most indoor hydroponic gardens require less than twenty hours of maintenance each week. And you will be growing the foods that you want to eat. And with the indoor hydroponic garden, you will be able to grow your plants all year round, no matter what the weather is doing outside

This type of gardening is environmentally friendly. The indoor hydroponic garden will not require that you clear land or allow the soil to erode when you cut all of the trees down to make a flat space for your garden. You will not need to worry about weeds and pests, so there will be no need to use potentially toxic pesticides and herbicides.

And you will be growing your produce locally and not depending on someone bringing it into you. You will always be in complete control of the climate. Just like growing plants in a greenhouse, the temperature indoors can be carefully regulated to ensure that your hydroponically grown plants will always have the perfect climate. And having the ideal environment means that you will be able to grow plants all year round.

Hydroponic gardening will save you time and effort. You will not be growing the plants in the soil so that you will not need to prepare the soil, remove the weeds by cultivating the earth, fumigate for weeds and pests, or water the plants regularly. This will mean that you will be able to spend more time enjoying your harvest and less time working on getting to your harvest.

Your plants will grow larger and healthier. Plants just naturally grow faster and better in water than they do in soil. Plants that grow in water do not need to send their roots digging through the soil in search of nutrients, so they are able to spend all of their time and effort on growing big and strong. Less water is used in a hydroponic system than in traditional soil gardening. The average water use of a hydroponic system is ten percent of the water that conventional soil gardening uses.

In the soil, most of the water either wash away or evaporates into the air. In hydroponics, the same water is used continuously as it is recirculated and reused. A minimal amount of water will be lost through evaporation. And when the fruits of your labor are harvested, you will not need to use water to wash the soil off of them.

You will be able to fit your hydroponic garden into any available space that you have. Even people who live in a small apartment will be able to set up a hydroponic garden. Since the plants will not need to have ample space for their root systems to spread out, you will be able to locate the plants closer together than you would be able to in a traditional soil garden.

You will have complete control over the nutrients that your plants will be eating. In the hydroponic system, the nutrients are mixed directly into the water that flows over the roots of the plants that you planted. This means that the roots will always have full access to the food that they need.

The pH level of the water mixture will always be within your control. You will have the ability to mix, measure, and adjust the pH levels any time you feel that they need to be changed.

There will be no need to guard against the traditional things that threaten the conventional soil garden. With your indoor garden, there will be no birds, bugs, gophers, or groundhogs waiting to attack your garden. And there will be no need to spend time pulling weeds because there won't be any. And the yield that you harvest from your hydroponic garden will be more stable and predictable than soil gardening.

And even if you are not planning to grow tons of fresh foods for your own use, gardening hydroponically is a wonderful stress-relieving hobby. Having this interest will give you a touch with nature. When you are feeling worn out from the day or need some time to destress, working on your little home garden will relax you tremendously. Since you do not need a lot of space to make a hydroponic garden, then almost anyone can have one in their home somewhere.

Disadvantages of Hydroponic Gardening

Like everything in life, there are good sides and bad sides to hydroponic gardening, and it is important to know all the details of both sides so that you will be properly prepared to overcome any obstacles that might get in your way. You will need to make some sort of initial investment in order to set up your system initially. You can plan to spend between fifty dollars and several hundred dollars, depending on how elaborate you want your system to be.

No matter which type of system you intend to set up, you will need nutrients, growing media, one or more timers, grow lights, and container, not to mention plants. Once you have set up your system, then you will only need to spend money on electricity and nutrients.

A certain amount of safety is required when operating a hydroponic system. You will be using electricity and water in the same place, so you will need to be careful to keep them separated. You should always think of safety first, especially when working with electric equipment and running water. And since you will depend on electricity to run your hydroponic system, you may need to have some sort of backup system in place, especially for the larger-sized systems. Without this, your system will stop working immediately if the power does go out, and if it is out for an extended period of time, then your plants might die.

While pests and diseases are rare in the hydroponic garden, they are possible, and they will spread quickly. Since your garden will be located in a small space, there will be less space for the pests to cover, and they will be able to spread quickly. Infections in the plants can also spread quickly because a hydroponic system is a closed system that uses water. When many plants are sharing the same reservoir of water, it is easy for one plant to affect another.

This problem is only a complication for the large, commercial warehouses. In a small home system, it is relatively easy to prevent the disease from spreading by having a sound system for managing the cleanliness of your system.

If eating only organic foods is important to you, keep in mind that hydroponically-grown food has not yet been determined to be organic. When produce is grown organically, it means that it is grown without the use of any type of genetic modification or fertilizers or pesticides. There are some growing methods available for hydroponic gardening that are organic in nature. Some growers will use particular growing mediums and organic nutrients.

While hydroponic gardening is not difficult, it will require some measure of commitment and time. Putting effort into your garden will give you the rewards of harvests that you want. The hydroponic system is different from the soil system in that plants that are grown in soil will survive when left on their own. They may not grow as lush and fruitful as you would hope, but they can still grow unattended. This tactic will not work when you have a hydroponics garden. Without adequate knowledge and proper care, your plants will die. Your plants are depending on you for their very survival, and you must take good care of them if you want them to thrive.

You will also need to take care of the system the plants are growing in. Eventually, you might want to automate the entire system, but in the beginning, it is best to be as manual as possible so that you can learn about your system and your plants. Even when your system is fully automated, you will still need to perform regular maintenance and cleaning. And along with this is the fact that you are using several types of equipment to run one system. You will need to learn as much as possible and keep studying and learning about your garden.

Hydroponics: Yes or No

Even after reviewing all of the disadvantages of hydroponics, do the advantages outweigh them enough to make hydroponics a lucrative adventure? Absolutely! There will be downsides to hydroponics, just like there are for many endeavors in life. Most of the disadvantages associated with hydroponics can be easily overcome with experience and planning. And hydroponics gardening has great future potential for growth and expansion. But even if you stick to your small home garden and concentrate on growing food for your own kitchen, it will be well worth it for you.

CHAPTER 3: TYPES OF HYDROPONIC SYSTEMS

A hydroponic system can be either passive or active. In a passive system, the system relies on the growing medium or a wick to move water and nutrients around through the system. In an active system, the nutrients and water will be moved around with the use of a pump. The system is also characterized as being either a non-recovery system or a recovery system. In a non-recovery system, the nutrient solution is applied directly to the growing medium, so it is only good for one use. In a recovery type of system, the nutrient solution will be re-circulated and reused.

A hydroponic system can be set up in one of six basic ways. When planting in a system, your plants will need oxygen, nutrients, and water or moisture. The biggest difference in the six methods of the building is in the way that these three things are provided to the roots of the plants. The type of system you will choose will depend greatly on the plants you want to grow and how comfortable you are in your own abilities. It is okay to start small and grow bigger with time and experience.

Drip System

This is one of the most widely used of all of the hydroponic systems. This system has very few working parts and is a completely straightforward method of hydroponic gardening. It is also very practical and versatile. Building a drip system will allow you to use your imagination in many ways. The way the drip system works is exactly the way that it sounds because you will be keeping the roots of your plants moist and well-fed by dripping a nutrient-filled water solution on them. You can design the drip system in many different configurations and in all sizes from small to large. The drip system is especially useful when you are growing large plants that will require ample room for their root systems. The lines for the drip system will easily run over a large space, so you will not need to use large volumes of water to flood the system. And the larger plants will use more significant amounts of the growing medium, which will help to keep the moisture gathered around the roots. This will help them to not become overly stressed if sometimes they are not watered exactly on schedule.

It is a simple operation to run a drip system. The appropriate nutrient solution is dissolved in the water, and then the water is pumped through the tubes out of the reservoir and onto the top side of the growing medium.

When the water reaches that point, it will drop out of the tubes and onto the growing medium. Then the water will soak the growing media and drip onto the roots, and then it will run back to the bottom of the container and back into the reservoir. It will flow back down into the reservoir at that point. The top of the reservoir will need to be positioned at least seven inches below the bottom of the growing tray so that the water will be able to run back downhill. And it will help to provide enough space for the larger plants that will need more room since they will have more growing medium and will not become stressed if watering is delayed.

There are two versions of the drip system. The recirculating drip system is the most common version of the drip system of hydroponics. This system does exactly what the name says: after the nutrient solution moistens the roots and flows back down into the reservoir, then it is reused. This solution can be used over and over again. The pH system of the recirculating system is subject to many changes, so you will need to test the pH of the water often in order to keep the water at the pH balance that is right for the plants that you are growing. And in order to keep the level of nutrients for the plants readily available, then you will need to change the nutrient solution regularly. Commercial growers are particularly fond of the non-recirculating system.

It might sound very wasteful to use the nutrient solution only once and then discard it, but that is not how the system is meant to work. With this system, you will water at precise times and with precise amounts, so that you will not use any more nutrient solution than you need at every feeding. None of the nutrient solution is wasted with this system and none of it is saved. And since the water is not sitting in the reservoir, it will not need to be cleaned as often as the recirculating system will.

One of the most important things that a drip system will need is a timed schedule for watering. The pump will be operated with the use of a timer, and the system will feed the plants several times each day. When the proper time is reached the timer will activate the pump and the pump will send nutrients and water into the container with the plants. Using a timer is critical for the life and health of the plants. In the drip system, if the flow of nutrients and water are left uncontrolled the roots of the plant will drown, and then the plants will die. If you choose to go completely manual and not use a timer for the pump, then you will need to set a schedule for manual operation of the pump to water and feed the plants.

In the hydroponic drip system, the best plants to grow are all types of lettuce, leeks, onions, and green peas, all types of melons, radishes, cucumbers, tomatoes, zucchinis, strawberries, and pumpkins.

Flood and Drain System (EBB and Flow)

The flood and drain system is not quite as simple as the wick system, but for setup and maintenance, it is an easier system than some of the other systems are. The flood and drain system involves using the nutrient solution to periodically flood the plants in the system, and then to let the water drain back out. The nutrient-filled water solution will flood into the growing container, where it will thoroughly soak the roots of the plants, and then the solution will drain back into the reservoir. A timer controls the cycle for pumping the water through the system. When the timer goes off, it will trigger the pump to begin pumping the nutrients and the water in the reservoir. They will then flow upwards through vinyl tubes into the plant container, where they will flood the plants until the roots have taken in all of the solution that they can hold. Then the rest of the solution will return to the reservoir through return vinyl tubes until the leftover water is all back in the reservoir. The system has an overflow tube that will keep the water from spilling out of the reservoir while the water is circulating for feeding time.

When the timer stops, all of the remaining solutions will drain back into the reservoir. The flood and drain system is very versatile, and you can assemble one for the very little initial cost. When you want to add or remove plants from this system, you will be able to do so without disturbing any of the other plants in the container. While it may seem rather complicated to set up and operate all of the different components of the flood and drain system, you will use a minimal amount of electricity and water, the system itself will produce plants efficiently, and it is relatively effortless to maintain. The plant tray, the reservoir, and the submersible pump with a timer are the three major components of the flood and drain system.

The plant tray is also known as the flood tray or the flood container. This is a large shallow container that you will use to put your plants in. Then you will add a growing medium to the container itself and place your plants in netted pots or pots with perforations, and then you will set the pots into the growing medium in the tray. The pots will need to be about twice as deep as the flood tray is. When the water comes into the flood tray, it will flow up into the bottom of the pots and feed and water the roots of the plants. Then the water will drain back into the reservoir, giving the plants time to dry out a bit so that the roots can collect much-needed oxygen.

The nutrient reservoir will need to be placed almost right below the stand that the flood tray is resting on. You will use a drain tube and a fill tube to connect the reservoir to the flood tray. Then the submersible pump with the timer attached will be connected to the fill tube, and this is what will control the amount of water that will flow up and flood the plants in the flood tray. The water will be pulled back through the drain tube and into the reservoir through the force of gravity. You will be able to use the same water in the flood and drain system for one week before you will need to discard it and make the fresh solution, to ensure that the plants are always receiving the proper amount of nutrients. Using the timer on the pump will allow you to control how much solution is used at each feeding, and how long each feeding will last. And you will be able to grow virtually any plant with this system as long as the flood tray is large enough to contain it.

You will need to implement a regular schedule for cleaning and sterilizing the components of the flood and drain system, since water will be continuously sitting in the reservoir and flowing into the flood tray. Some of the harmless forms of algae will grow in the flood tray and the reservoir, and this is quite normal. But if the components of this system are not cleaned regularly the mold growth and insect infestations will build up, and they will kill your plants.

Whatever growing medium that you use will need to have some weight to it so that the plants do not float away when the nutrient water floods them. The perfect growing medium for this system will be able to retain some of the moisture when the water is gone but will also drain well. When you are growing root vegetables, they will need to be able to push the growing medium off to the side so that the roots will have room to grow. So the growing medium will need to have weight but also be able to move freely.

The best plants to grow in the flood and drain system are plants with deep root systems like beets, chicory, and comfrey, or those plants that will need to be staked to a trellis: like cucumbers, squash, tomatoes, and beans.

Wick System

When you are just beginning your journey into hydroponics gardening, you will not find any system that is easier to set up and operate than the wick system. This system requires the use of only four separate parts. If you are a fan of upcycling or recycling then this might be the perfect system for you, because you can use whatever you have laying around, or thrift store bargains, for the components of the system. Probably the most difficult decision that you will need to make is what kind of material to use for your wicks, since you will have so many different choices available to you.

This system uses a wicking action to get the nutrient solution from the reservoir and up to the plants that are growing above. The four components of the wick system are the container or tray to put the plants in, the holding reservoir for the nutrient solution, the growing medium for the plants, and the wicks themselves. The container that you use to put the plants in will be located slightly above the reservoir for the nutrient solution. These two containers will be connected by the wicks. The wicks will bring the nutrient-filled water up into the tray where the plants are growing and into the growing medium to feed and water the plants. The water will be absorbed by the growing medium, and the plants will be fed and watered.

Capillary action is used to bring the nutrient solution up the wick and into the growing medium. This action is nothing more than the ability of a liquid to move upward and against the flow of gravity. The way that the wick of a candle draws the wax upward or the wick of an oil lamp draws the oil upward are both examples of capillary action. This system of hydroponics is a passive system because it works without the use of any kind of motor, pump, or moving part. This does not mean that you can't use a pump in the reservoir to keep the solution flowing, as this will keep it fresh, and some gardeners prefer to do this. It just means that no moving parts are required to make the system work properly.

If your hydroponics garden is located in an area that is full of natural light, then you will not need to make use of a grow light. This fact makes the wick system the most environmentally friendly of all of the hydroponics systems. You can use recycled items and renewable materials for the components of your system. And the wick system of hydroponics will use the least amount of water.

The growing container that you use for your wick system can be made from anything available, like a plastic tub, a bucket, an empty soft drink bottle, or a custom grow tray specially made for the wick system. You will put the growing medium into this part of the system, and then you can either set the plants all together in the growing medium, or you can set them in the medium in their own individual pots. There will need to be small slits or holes in the bottom of the growing tray so that you can pass the wicks from the reservoir and into the growing tray. Try not to make the container for the growing medium too complicated because you will need to be able to clean out the container periodically.

You will not need to cover the reservoir, but it will need to be of some material that will not let light through, or it will need to be spray painted so that the light will not shine through. Algae and mold require food, water, and light in order to be able to grow.

If you remove the light from the equation, then the reservoir will not fill with growing mold and algae. You will need to place the reservoir as close to the bottom of the growing tray as you possibly can, so that the nutrient water will easily be able to flow up the wicks and into the growing container. And you will need to be able to get to the reservoir easily so that you can refill it and clean it as needed. You can install a pump in the reservoir to keep the nutrient water circulating and aerated, but this is not needed.

For the wick system, you will need to choose a growing medium that is superb at absorbing and holding quantities of moisture. This is how the plants will be fed, from the moisture in the growing medium, and so it will need to be an absorbent one. You can use any type of substance that is shaped like a rope to make the wick out of. It will need to be absorbent enough to be able to draw the nutrient water u from the reservoir and into the growing medium. Wicks can be made from cotton, wool, felt, nylon, string, yarn, or rope. Some of the more popular things to use for wicks are torch wick, strips of felt, microfiber cloths, or terrycloth toweling. You will need to thoroughly wash the material before using it for a wick so that you can remove any chemicals that might be on the wick material. Experiment with different kinds of material for your wicks.

The most important consideration for the wick material is that it be absorbent and sturdy so that it will bring up water and not rot in the process. You will want to use two wicks for each plant.

There are numerous advantages to the wick system of hydroponics gardening. They are efficient in their use of water and nutrients, they do not require electricity, they can be made from recycled materials, they are easy to maintain, and they are simple to build. The biggest disadvantage to the wick system is that the growing medium will need to be cleaned periodically so that it will not build up a level of nutrients that will be toxic to the plants. The wick system will also put some limits on the kinds of plants that you will be able to grow. You will not be able to grow plants that require a lot of water, like fruits, or plants that grow to be overly large.

And even though it is a low maintenance kind of system, the wick system will still require regular maintaining. You may be concerned that the plants are not getting enough water and nutrients because the wick system will deliver these to the roots very slowly. As long as you make sure that the growing medium is consistently moist and that your plants do not look thirsty, and then you can be confident that your system is working the way that it should be.

Plants that are growing in drier air or higher temperatures will need more moisture, as well as larger plants. You can always change to a more porous growing medium or add in more wicks if you feel that your plants need the extra moisture. You will probably not need to worry about your plants getting too much water, but if they show signs of being over-watered, then they are not getting enough oxygen, and you can reduce the number of wicks to allow the growing medium to dry more between feedings. To prevent the nutrients from building up in the growing medium, you will need to rinse it thoroughly every two weeks minimally. If nutrients are allowed to build up in the growing medium, they can have a toxic effect on your plants. This will show itself in smaller root masses, weak plant stems, slower growth of the plants, lesions on the plant, or premature loss of leaves.

Aeroponic System

With the aeroponics form of hydroponics gardening, you will grow your plants in a misty environment without the use of any form of soil or growing medium. Aeroponics builds off of the basic idea of hydroponics because the plants are grown in water but without a growing medium. The roots of the plants will hang in the air, where they will periodically be misted with a nutrient-rich solution. You will plant the seeds of your plants in tiny pieces of foam that are then stuffed into small pots.

Then the pots that are holding the seeds will be exposed to light on one side, and a nutrient mist on the other side. The foam in the pot will hold the plant in place while it is growing.

This system will require the use of some sort of enclosure to keep the light from reaching the roots of the plants, and to hold in the humidity around the plants. You will also need to have a separate tank to hold the nutrient solution that will be used to make the mist. As long as all of these components are used, then you will have some wiggle room for exactly how the components of the system are positioned. The aeroponics system can be arranged vertically or horizontally. When the system is set up vertically, the roots will have ample room to spread out and grow, while the system itself will not take up much space. Also, with the vertical system, you can put the misting devices at the top of the system, so that the mist will be allowed to flow over all of the plants from the pull of gravity.

You will also need to decide if you plan to build a high-pressure aeroponics system or a low-pressure aeroponics system. The high-pressure system will provide a true mist for the plants, because it will be able to provide the high water pressure that is required. This requires the use of more than one pump to build up enough pressure.

The low-pressure aeroponics system is sometimes referred to as a soakaponics system because the system relies on the water to be sprayed through a simple fountain pump. This process creates a gentle spray that is not an actual mist. No matter which system you use, you will need to have extra sprinkler heads available for the system because they will sometimes become clogged from the nutrients that are in the water.

Theoretically, you can grow any type of plant in the aeroponics system. This system is generally used for culinary herbs, strawberries, cucumbers, tomatoes, and leafy salad greens. The one type of plant that does not do well in the aeroponics system is any type of root crop, because the actual part of the plant you will be eating will be soaking in water constantly, and will be more likely to rot than to grow. And because of their size, trees and shrubs usually do not do well in the aeroponics system.

Aeroponics systems use water efficiently, and they grow plants that not only survive, but they seem to thrive in a misty environment with their roots dangling in the air. Part of the reason that this system works so well is that the roots of the plants are able to soak in extra oxygen, and this will give you faster plant growth.

And you will be able to grow a larger quantity of food in a smaller space, especially if you set your system up vertically. And there is no need to worry about water escaping from the system because this system is completely enclosed. The aeroponics system will need a bit more maintenance than some of the other systems will to keep it operating at maximum efficiency. You will need to carefully maintain the level of nutrients in the water. If your misters are not working correctly, then your plant roots might dry out, and this would result in the death of the plant. And you will need a constant flow of electricity to make the mister work properly.

Deep Water Culture System

This system is also known as the bubble bucket system because of the manner in which the simplest form of the system can be built. All you need for the simple form is a five-gallon bucket that has an attachable lid. Then you will cut a hole in the lid that you will set the pot with the plant into. The pot will be made of netting, so the roots have access to the nutrient water. Then you will fill the bucket with nutrient water almost all of the way to the top, leaving the top half of the plant roots out of the solution so that they can collect oxygen.

You will place an air stone or a submersible pump in the bucket to create air bubbles to provide oxygen to the roots, and then you will have a fully functional bubble bucket system. The only system that is easier to set up than the deep water culture system is the wick system.

You will not need to use any kind of growing medium in the net pots because the roots of the plants are meant to be sitting partially in the water. However, it is a good idea to use some sort of growing medium in order to keep the plants in the pots from floating up and out of the nutrient water. This will also prevent the plants from tipping over in the pot if they do not fully fill the pot. You can use small river rocks or aquarium pebbles for this purpose.

The fact that the plants have continuous access to an unending supply of oxygen, water, and nutrients is one of the things that make the deep water culture system of hydroponics so successful. The water solution needs never cover the whole root because the root will also need to collect oxygen in order to be able to grow properly. You will probably use the air stone or submersible pump in order to make the oxygen that the plant will need. If you can use a larger-sized air stone, that will be the better option because a larger stone will create more air bubbles and this will make more oxygen for the plant.

The simplicity of the deep water culture system is the best part of it. All you will need to set up the system is the air stone with the pump, a tray for the plants, and a reservoir for the water and nutrient solution. Plants will grow fairly rapidly in this system. This means that you would be able to harvest a head of lettuce in half of the sixty-day time period that is required for growing in soil. This system has very few moving parts and is relatively easy to assemble. While the deep water culture system is a simple form of system, it will need some regular attention and can't just be set up and then ignored. If the system is not monitored regularly, then the level of the water, the pH level of the water, and the concentration of nutrients in the water can wildly fluctuate, and this can be damaging to the plants, and possibly fatal. You could very easily over-feed or under-feed the plants. If the electricity goes out and the pump is not pumping bubbles, then the roots of the plant might drown. And you might have problems with maintaining the temperature of the water at an even level.

In this system, you will want to grow plants that do not need to produce flowers in order to make a crop. You will be able to grow almost any type of lettuce and most herbs in the deep water culture system. These plants will grow quickly and quite well. You may be able to grow some of the plants that require flowering, like tomatoes, peppers, and squash; these plants will just require an extra bit of work.

Nutrient Film Technique System

This system is an extremely popular system with both commercial gardeners and gardeners who grow at home on a smaller scale. The reason that the nutrient film technique (NFT) system is so popular is that its basic design is quite simple. All different types of greens will grow well and quickly in the NFT system, and they are well-suited for this type of system. All NFT systems have some aspects of their design in common, even though they are able to have many different designs. Every NFT system will use individual pots in which to hold the plants that are being grown. Since the roots of the plants will be able to gather enough moisture, nutrients, and oxygen from the water that flows through the system, many growers will not use any type of growing medium in the pots. The netted pots will sit in a container called a growing tray, and this tray will be set slightly lower at one end than it is at the other. This configuration will allow the water to flow into the tray at the higher end and flow back out on the lower end. The upper part of the root remains dry with this system; so that they are able to absorb oxygen from the air while the lower half of the roots collect nutrients from the water solution that flows over them. The reservoir will hold the nutrient water solution in the NFT system. A large tube will be used to pump the nutrient solution up to the growing tray.

The solution will be pumped into a tube manifold, and from there, it will be pushed through several smaller tubes that will run into channels on the growing tray where the plants are located. Since the channel is sloped slightly, the nutrient solution will run into the tray at one end, through the channel where it will flood the plants, and then out the other end and back to the reservoir. Suspended above the feeding channel are the individual plants in their individual pots, with their roots hanging into the solution just enough so that they can collect nutrients as the solution flows past. The water will flow past the plants in a shallow film so that the roots can soak up the correct amounts of nutrients that they need. The amount that is excess will flow out the other end of the channel, back to the reservoir where it will be used the next time.

Two of the most important considerations to follow with the NFT system are how deep the water will need to be and how fast the water will flow in the channel. The angle at which the growing chamber is sloped will determine just how fast the water will flow through the channel. If the tilt is deeper than it should be, then the water will flow out too quickly, and the plants won't receive enough nutrients. A shorter feeding tube is preferable when you are building an NFT system. You will need to have one inch of drop (slope) for every thirty inches of length of the feeding tube.

To put this in simpler terms, if you are using a thirty-inch long feeding channel, then one end of the channel will be one inch lower than the other end. The channel will also need to be as level as possible, because any bit of sag or droop in the channel will allow water to pool there and possible damage the plants.

The number of plants being grown at any one time and the size of the plants will determine the amount of water that you will need for the system. Five gallons of water in the reservoir will allow you to grow between forty and fifty plants. You will need to add one more gallon of water for every additional twenty plants that you want to add to your garden. If you are growing plants from seeds, then you will use just half of the amount of water while the plants are becoming seedlings. A plant is a seedling when it has two real leaves. As the plants grow larger, don't forget to increase the amount of water and nutrients that you have in the reservoir. The water pump will run constantly in the NFT system, and it will not be put on a timer. You will need to check on the pump often because if it fails, then you could lose your whole garden.

Any kind of leafy greens will thrive in the NFT system, like spinach, arugula, and kale, or any type of lettuce green like iceberg, Bibb, and romaine. You will also be able to grow strawberries and almost any type of herb.

No matter which form of hydroponic system you choose to assemble, you will find that the system has its own advantages and disadvantages. You will need to decide if the advantages of the system that you choose to assemble outweigh the possible disadvantages of that system. And there is no rule that says that you must stay with one type of system because you should experiment with all of the forms to see which one works best for you.

CHAPTER 4: MATERIALS FOR BUILDING A HYDROPONIC SYSTEM

The hydroponic garden will allow you to put a lot of garden space into a small amount of physical space Since you will be bringing the nutrients to the plants as opposed to putting the plants in the soil and making them search for their own nutrients, The plants will not need as much space to grow, and they will grow more quickly. The system you will build for your hydroponics garden will depend on how much area you have for your garden and how many plants you want to grow. This last part is important because certain plants will grow better in certain systems. You will also need to think about whether or not you will ever need to move the system and how much cleaning and maintenance the system will require. You will also need to know and understand the different components that you will be using to create your hydroponics system so that you will make the best possible choices for your system. While the different systems will do different things for your garden, they will all use the same basic components that you will use to build and maintain your system.

Grow Lights

The grow light is something of an optional piece of equipment for the home hydroponic system. While your plants will need a source of light, if they are exposed to a proper amount of natural light, then you will not need to use grow lights. Natural light is free, so it is the best choice, but you will often need the boost that artificial lighting will give to your garden. If your area does not get a lot of sunlight, or if you can't assemble your system in an area of natural light, then you will need to set up grow lights.

Plant grow lights are not the same thing as regular household lights. The grow light has a bulb that is specially designed to mimic natural sunlight by putting out certain spectrums of colors. The plants need light to create the process known as photosynthesis, which is the process the plant uses to provide food for itself that is independent of the nutrients that you will provide. Your plant's growth will depend greatly on its ability to complete this process on a regular basis.

You can choose from various forms of grow lights to use with your system. Each style will have its own unique advantages. You will want to use a compact fluorescent light if your hydroponic garden needs to be set up in a small space, because these lights do not produce much heat or use much electricity.

They are also able to be attached to a wall and set sideways in order to be used with a vertical system or for taller plants. You will use a high output fluorescent light if you want to increase the amount of light that shines on the plants or if you want to move the lights to be closer to the plants. If you use LED lights, then you will grow thick, strong plants. These plants will have leaves that are a bit thicker than normal leaves, and the plant will progress normally while it produces a healthy harvest that will be a bit slower than most harvests. The high-pressure sodium bulb will have its strength in the yellow and red regions of the light spectrum which is the area of light that fruiting plants and flowering plants seem to prefer. The grow light that is the closest to natural sunlight is the metal halide bulbs, which are strong in the blue end of the light spectrum.

Submersible Pump

A submersible pump will be used in any system that requires a pump to pump the solution of nutrients and water up to the area where the plants will be growing. You will be able to find good submersible pumps in any home improvement store or garden supply store. These pumps will come in many different sizes, so you will be able to find a pump that will fit the needs of any hydroponic system. The idea behind the submersible pump is that the impeller will be spun around with the use of an electromagnet.

You will be able to easily take the submersible pump apart to clean it. You can use a small bit of furnace filter to keep the submersible pump free of debris if it does not come with its own filter. And the pump and the filter will both need to be cleaned regularly in order to prevent the growth of mold and algae.

The size of the pump that you will need to use will be determined by the size and configuration of our hydroponic system. There will be slight differences in the pump size based on the size of your system. The volume of water that the pump will be pumping will determine the size that it will need to be, so you will need to know the minimum number of gallons of water that you will need to pump through your system. You will always have the ability to reduce the amount of water that the pump is sending through your system, but you will never be able to increase the volume if your pump is not big enough.

Reservoir

Because you will not be able to have a hydroponics garden without a reservoir, this will be one of the most important pieces of your design. The reservoir is the component that you will use to hold the water and the nutrients. The nutrients will be dissolved in the water, and the water will be used to water the plants that you are growing.

Whatever type of hydroponic system that you are assembling will determine how the reservoir will be used. The reservoir can be used as a growing chamber if the plants are allowed to hang their roots in the solution of the reservoir, or the reservoir will hold the nutrient solution until it is pumped up to the growing chamber above.

You can use almost any container for the reservoir as long as it is clean, holds water, and keeps out the light. You can easily use recycled items as long as they are cleaned well first. One of the most important things is that the container either be lightproof, or that you are able to make it lightproof. When you hold the empty container over your head and face a strong light, if you can see light through the sides of the container, then it is not light proof. The easiest way to make any container lightproof is to spray it only on the outside with black spray paint that is made for plastic, and then cover the black spray paint with glossy white spray paint. The black paint will keep out the light, and the white paint will repel the light. The light must not be allowed to enter the reservoir to discourage the growth of mold and algae.

One of the most basic problems that new gardeners face is not choosing a reservoir that is large enough for the types of plants or the number of plants that they want to grow. Plants will begin small, but they will get bigger over time.

Those plants will need to have access to more water and more nutrients. You can use a simple test to ensure that the reservoir that you choose is large enough for your needs. No matter whichever system you are building, the ratio of water per plant and the nutrient solution will always remain the same. Find out what will be the average size of the plants when they have reached their full growth. Once you know how many plants you will be growing you can determine the size of the reservoir you will need.

Growing Tray

The part of the hydroponics system where the plants will be located is known as the growing tray. It can also be called the growing chamber, flood tray, flood chamber, or botanical tray. This tray will need to be sturdy enough to hold the plants and the growing medium, as well as any water that might be in the tray. The tray will give the plants access to the water and nutrients as well as providing support for the plants. The tray will also be responsible for protecting the plant from heat and pests. Keeping the plants cool is important, because excess heat will stress the plants and damage the roots, and may even result in the death of the plant. And while the plants will need light to grow, they also need periods of darkness, so they will need to be protected from excessive amounts of light, which the growing tray will help with.

If the roots of the plant are damaged and it is a flowering plant, the flowers may fall off, which will cause the crop to not grow since it needs the flowers to provide the fruit.

The shape and size of the growing tray will be determined by the kind of plants that you choose to grow and the size of the system you are assembling. Larger plants with a larger root system will obviously need a larger tray. There are only a few types of containers that you would not want to use for your growing tray. Using any container that is made out of metal or coated with some type of metal is not a good choice for a growing container. The metal might rust in the water or the nutrients might react unfavorably to the metal. And if the metal flakes off it can get caught in the vinyl feed tubes. Don't be afraid to use recycled materials for your growing tube as long as they are easy to clean and sturdy.

System for Nutrient Delivery

The plants in the growing tray are fed with the water in the reservoir that nutrients have been added to. This is really a very simple system to assemble and operate. When you are building your own system, you will be able to customize the delivery system to fit your own needs. The submersible pump, the drain tube in the growing tray, and the vinyl tube that delivers the water to the growing tray are the parts of the nutrient delivery system.

It is best to use PVC tubing and connectors because they will not break down or collect mold. The vinyl tubes that are used to carry the nutrient water from the reservoir to the plants should be either black or blue in color; black is the preferred color but blue is also acceptable. This will prevent excess light from getting through the tubes to the nutrient solution. Excess light will cause mold or algae to grow, degrade the nutrients as they pass through, or allow the nutrients to become too warm where they might damage the roots of the plants. The pipes, tubing, and connectors can be the same as the parts that are used for sprinkler systems or outdoor water features, and they can likely be found in any plumbing supply store or big box retailer. If your system design calls for them, then you will also need drip emitters or sprayers, and if you use these, make sure you keep extras on hand because the nutrients passing through do tend to clog these.

Air Pump

The deep water culture system is the only one of the hydroponics systems that require the use of an air pump. This accessory is an optional item in all of the other system configurations. But there are definite benefits to using an air pump with all of the systems and, since they are relatively inexpensive, it makes no sense to not use one. Using an air pump in the reservoir will help to provide fresh air and oxygen to the roots of the plants.

The air pump will pump air to the air stone in the reservoir and this will help the system to provide oxygen to the plant's roots. And the bubbles that are moving around in the reservoir will help to deter the growth of mold and algae by keeping the water solution fresher.

Growing Mediums

The use of a growing medium is not required in all of the systems but it is in some. The growing medium will be used to support the plants while they are growing and also to provide the plants with nutrients. Hydroponic gardening requires the use of special types of growing mediums. The growing medium is made of an inert material that will support the plant and hold it in place without providing any type of nutrients to the plant. There are many different types of growing mediums that are available today for the hydroponic garden. No one type of growing medium is better than another type unless you are looking for a growing medium for a specific purpose.

Coco coir – This is a completely organic substance that is created from the empty shells of coconuts. This growing material is available in different consistencies. It retains water well and has excellent drainage. Coco coir is also a great choice for germinating plants from seeds.

One of its best advantages is that the coco coir is completely renewable and sustainable. Coco coir, also known as coco chips or coco fiber, breaks down quite slowly and is pH neutral, and it also provides good aeration for the roots of your plants. Coco fiber is comparable in size to potting soil, while the coco chips are more like the size of small wood chips. The chips are good to use if you are using baskets to grow your plants in, because the chips are too large to fall out of the slats or holes in the baskets.

Vermiculite – This growing medium is made out of a silicate material that expands whenever it is exposed to very high temperatures. When used as a growing medium, vermiculite tends to float and is very lightweight, and it will hold onto nutrients for feeding the plants later. When it is used in hydroponics, vermiculite is often mixed with perlite to assist with drainage, since vermiculite tends to hold water and perlite lets water drain out easily. When it is used in hotter climates vermiculite is often used alone. And since vermiculite tends to break down easily and quickly, it is best used for plants that have a short growing period like lettuce does.

Perlite – This growing medium can be used by itself or with other types of growing mediums. Perlite is made from various minerals that are exposed to a high heat, which makes the pieces become absorbent, porous, light weight, and expanded, much like popcorn. It has an excellent capacity for wicking and is pH neutral. Perlite used by itself is not the best choice for all hydroponics systems because it is so lightweight that it will float.

Rockwool – This growing medium is one of the most common that is used in hydroponic gardening. It is a non-degradable medium that is porous and sterile. Rockwool is made from limestone and/or granite that is heated to a high temperature and then melted, and then the melted liquid is spun into threads in a process that looks like making cotton candy. The threads can then be shaped into flocking, slabs, cubes, sheets, or blocks. Rockwool likes to suck water up and hold it, so you will need to be careful not to let it become saturated because it could rot the roots of the plants. You will need to soak the rockwool in pH balanced water before using because it is not naturally pH balanced.

Hydrocorn (grow rock) – This is a substance that is known as a Lightweight Expanded Clay Aggregate, which is a type of clay that is fired to a super high temperature, which gives it a porous texture.

The resulting material is lightweight but also heavy enough so that it will provide secure support for your plants. Hydrocorn is pH neutral, sterile, and non-degradable. It is reusable and very good at wicking up moisture for your plants. Since it is reusable, it will only need cleaning and sterilizing between crops.

Floral foam (hydroponic foam) – This is the same substance that you find in the bottom of the vase when you buy a vase full of plants. It is the foamy green substance that artificial plants are stuck into to make them stand upright. Floral foam will crumble easily, and that may cause it to release small particles that will get into your water lines. And floral foam will absorb water and hold it very well, so it might become waterlogged. Floral foam is best used as a medium for germinating plants from seeds.

Oasis cubes – These are similar to cubes of rockwool, and they have similar properties, but they are more like floral foam in their composition. The cells of the oasis cubes will readily absorb air and water. Oasis cubes are usually used as a medium for germinating your plants from seedling, and with some of the more delicate plants, the oasis cube can be moved directly to the growing tray, which eliminates the need for transplanting the more fragile plants.

River rock – This fairly inexpensive growing medium can be found in pet supply stores and also in home improvement stores. These rocks have smooth edges and they are rounded. They are not porous, so they will not hold moisture in the roots of the plants. Since river rocks are uneven in shape and size, they will naturally form pockets for air circulation. These air pockets will also allow water to drain out well. It is best to not use river rocks alone but to pair them with another growing medium like coco coir. The rocks are popular with growers who like to line the bottom of the growing tray with them and then laying another growing medium on top, which will allow the roots to remain moist without leaving them sitting in water.

Pine shavings – These are usually used in large-scale drip systems and the shavings should never be confused with sawdust. Pine shavings are inexpensive, but you will need to ensure that the ones you buy are made from wood that was dried in a kiln and has not been treated with fungicides. Drying the wood shavings in a kiln will kill off all of the sap in the wood that might harm the plants. Since pine shavings are made of wood, they will absorb water easily and may allow the roots to become waterlogged. The shavings will either need good drainage or they will need to be placed on top of river rock.

Water absorbing crystals – These are polymer crystals that are used in many different industries. You might know that they are used as the absorbent material inside of disposable diapers. They are also used in the cloth rags that athletes and body builders use to keep them cool. The crystals will soak up water and expand to many times their natural size. They are not currently used much in hydroponics, but their use is increasing all of the time.

Timer – Depending on the type and size of the hydroponic system you are assembling, you will need one or more simple timers. You will need a timer for the grow lights if you are using artificial light instead of natural light, to turn them on and off automatically. You will need a timer to control the submersible pump unless you want to leave it running all of the time. And you will need a timer to control when the nutrient system is pumped up to the plants. The light timer and the submersible pump timer can be simple on-and-off timers, but the timer for the feeding system will need to have several sets of nibs on it for several sets of on-and-off timing since the plants will need to be feed more than once a day.

Pots – When shopping for pots for your hydroponic system, you will need to consider which pots are the best options for the plants that you are growing.

Net pots -- also known as hydroponic pots, are the best type of pots to use for aeroponics systems, nutrient film technique systems, flood and drain systems, and deep water culture systems. These pots will support both cuttings and seedlings. Since the bottom of the pot is made from net material, then it will allow the roots to be aerated and the water to properly drain from the pot after feeding. The net area will also let the roots of the plant have better access to the nutrient water. They are reusable if they are properly cleaned and sterilized between uses. Use net pots with perlite, rockwool, grow stones, or hydrocorn.

Fabric pots -- These pots range in size from one gallon to five hundred gallons, although you will probably never need something that large for your home garden. Since the pots are made from fabric, they allow good aeration to the roots and proper draining of the nutrient water so the roots are not subject to rot. The best types of growing medium to use in these pots are rockwool, grow stones, hydroton, and coco coir.

Plastic pots – These pots have been in use by gardeners since plastic was invented. They have holes for drainage in the bottom and strong, sturdy sides.

When using plastic pots, always buy a pot to fit the size that the plant will be when fully grown. This will eliminate the need for constant transplanting. Any type of growing medium can be used with plastic pots.

Materials – There are several miscellaneous building materials that almost every design of hydroponic system will need for assembly. The configuration of your system will let you know what you will need buckets, plastic tubs, tubing, stoppers, fittings, connectors, regulators, drippers, and valves, emitters, and hose clamps.

What you will need to purchase will change with the configuration of your system and as you learn more about hydroponic gardening. You will be able to see what you need and what you will be able to use from materials that you have at home.

CHAPTER 5: NUTRIENTS, PEST CONTROL, AND PH BALANCING

While they may seem like small things, your success at hydroponics gardening can be made or broken by specific things like nutrients, pest control, and pH balancing. Your plants will not be able to get nutrients from the soil they sit in, because there is no soil. An indoor garden will not be as susceptible to attacks from pests, but there are still pests that might like to visit your garden. And since you will be using water as the medium for growing your plants, it is essential that the water you use has the proper pH balance. These three things will need to be attended to because they are likely responsible for at least fifty percent of the success of your garden.

Nutrients

Every plant will require a certain amount of fertilizer if it is to grow tall and robust and give you a good harvest. When plants grow in the soil, they will get most of the nutrients that they need from the land they grow in. But this is not the case with plants that are grown hydroponically, since water has no natural nutrients that the plants can use. The water that is used to water the plants needs to be supplemented with nutrients.

The types of nutrients that plants need are divided into two categories: macronutrients (macros) and micronutrients (micros). The nutrients which your plants need in large amounts are the macros, and those are calcium, magnesium, potassium, oxygen, hydrogen, carbon, sulfur, nitrogen, phosphorus. Micros are still essential nutrients, but the plant only needs these in small amounts. These are boron, manganese, copper, nickel, chlorine, molybdenum, iron, and zinc. Plants are not able to complete their life cycles, which includes completing enzymatic reactions and building molecules, without the proper mix of essential nutrients. In the word of hydroponic gardening, this would mean that your plants will not be able to produce the vegetables and fruits that you desire, or if they did, the harvest would be less than ideal.

The solution of nutrients in the reservoir will need to be kept at a steady, ideal temperature. The best temperature for the plants and the nutrient solution is somewhere between seventy and seventy-eight degrees. Many growers overlook the importance of temperature . When the temperature of your nutrient solution is just slightly outside of the optimal range, say two degrees, either way, this is not likely to harm your plants. It is relatively easy to check the temperature of the water in your reservoir. You will need to purchase a small glass thermometer that was designed for use in aquariums and attach it to the wall inside of your reservoir.

The ones that change colors are not the most accurate; instead, purchase one that has actual mercury in the thermometer. You can also keep the thermometer near the reservoir and just drop it in to test the temperature of the water once or twice a day. If you tie a string around some part of the thermometer, then it will be easy enough to drop it down into the water and pull it up again.

If the temperature of the water in the reservoir becomes too high, it will stress your plants during watering. The water will cause heat stress for the plants, and that will cause the plant to begin to shut down and go into survival mode. The plant will let you know when that is happening by dropping its flowers or leaves, curling up and turning brown, and turning to seed. The roots of the plants will turn black, become slime, and ultimately they will die.

Water temperatures that are higher than ideal will also create the perfect environment for damaging bacteria, fungi, and microorganisms to grow. These will multiply rapidly in the roots of the plants as well as in the reservoir. The good bacteria that would normally keep these things in check are not able to survive in the higher temperatures. And as the plants begin to die, they will emit toxins that will help to feed the harmful organisms.

If your nutrient solution stays at the proper temperature, and you use the correct mix of nutrients, then you should have no problem reaping an excellent harvest from your hydroponic garden. You can mix your own nutrients that are planned to the exact mix that your individual plants require, or you can purchase pre-made solutions. Home hydroponics growers usually buy a pre-made solution of a powder or liquid concentrate so they can just add the nutrients to the water and stir.

The pre-made concentrate solutions generally come in two separate bottles. One will be the micros, and one will be the macros, and they are kept separated because some of the ingredients are not compatible with other ingredients if they are mixed together as concentrates. Once the macros and micros are diluted, then they can safely be mixed together without issue. The nutrient solutions are easy to mix for the hydroponic system. You will need a measuring cup, a container to mix them in, and something to stir the mixture with if the container doesn't have a tight-fitting lid. If the container has a lid, then you can just shake the mixture until it is well-blended. If you are using a large scale flood and drain system, then you might need to mix up large quantities of nutrient solution at each mixing. If your system is smaller, then you will only need to mix the amount of solution that you will need for mixing into the reservoir.

After you mix your nutrient solution, you will want to check the pH balance of the solution and adjust it as is needed. If you start off with a perfectly balanced pH, then your solution will be easier to maintain.

Pest Control

As the owner of the home garden, there are certain things that you can do to reduce the chances that some sort of garden pest will make its home in your garden. Most of the practices that you can take advantage of involve a lot of avoidance and a little bit of knowledge. You should never go into your garden area if you or your clothing is dirty. There are all kinds of contaminants like pests and bacteria that can live on your clothing. It is better to take the few minutes needed to change clothes than to risk contaminating your garden space. The tools that you use in your garden are will also need to be scrupulously clean. Don't bring anything, including you, into the garden area if it isn't free of contaminants and completely clean.

You will want to clean everything that you are planning to use on your system before you use it. That means cleaning everything before you assemble the system, even the items that are brand new.

Even plastic tubes or hose clamps might have contaminants on their surfaces, not to mention how many different people had touched them before you bought them. And make sure that any windows or doors in the room are well-sealed so that they close tightly.

One of the most essential components of the hydroponic garden is also one of the most common things that will introduce pests into your garden, and that is the growing medium that you use. Most of the growing mediums that you will buy are sterile and perfectly safe, but there are some things you will want to look out for. If you are buying an organic growing medium like rice husk or coconut husk, they may be harboring pests that will wreck your garden. If you aren't sure that the growing medium you are buying is sterile, then you will need to sterilize it before using it. You can easily sterilize your growing medium in a solution that is four parts of hydrogen peroxide to six parts of water. Let the organic medium soak in the hydrogen peroxide solution for at least four eyes, and then rinse it thoroughly and allow it to dry. If you are purchasing a transplant for your garden, try to make sure that they are coming from a reputable nursery or grower. Plants from outside sources can easily carry pests, disease, fungi, and bacteria. Examine your plants carefully before placing them into your garden.

You will begin pest control by implementing measures that will prevent the development of pests in the first place. You can prevent developing a problem with pests by taking a few precautions. Some pests, like fungus gnats and spider mites, are attracted especially to excess moisture and low humidity. By keeping the humidity surrounding your system from getting below fifty percent, you will be able to prevent an infestation of mites and help to keep your plants healthy. And don't let the humidity get above seventy percent because then you will be attracting the pests like fungus gnats that like a moist environment.

Even if you are diligent in your home garden maintenance, you can still develop a pest problem. You will want to inspect your plants regularly and look for signs of pest infestation. There are certain signs that you will need to look for that will help to alert you to the presence of a pest problem.

- Some garden pests will leave a pattern of spots when they are present, either in black, brown, yellow, or white spots. If you see spots on your leaves, then look for deposits that the pests might have left on your leaves from feces or eggs. If you are able to rub the spots off, then it means that you have a pest problem.

- Some garden pests will literally suck the nutrients out of the leaf, causing it to turn yellow or yellowish-brown in color. You will find this discoloration around tiny holes in the leaf where the pests have been feeding.

- An actual hole in the leaf is a sign of a pest infestation. You might sometimes see small burns on a leaf where the leaf has gotten too close to a light or heat source. The holes that are left by pesky insects are very small in size and often have raised bumps around them.

Once you have noticed the signs of a pest infestation, then you will need to fix the problem immediately, if you want to have any hope of saving the unaffected plants. You will be able to remove some of the pests by making changes to the plant's environment or by removing the infected plants, while some of the pests that infect your garden can only be eliminated by using chemicals. There are also some more gentle methods that you can use to try to remove the pests from your garden:

- Using sticky traps might help you get rid of the pests. These work by trapping the pest in the sticky material and holding it there until it dies. This method can also help you to identify the pest since you will actually be able to see the bug.

So even if the sticky traps do not completely eliminate the pest, at least you will be able to see them to know how to get rid of them. You can also use sticky traps as a preventive measure to hopefully catch the bugs before they get in your garden.

- There are solutions that are available on the market today that are touted as perfect for killing the pests but will also kill your garden. If you purchase a commercially available chemical pest killer, just make sure that it is marked as being safe for plants. Some pest killers are made from natural sources like flowers and herbs. You might also consider planting some herbs or flowers that are natural pest repellants like the marigold that pests can't stand the smell of.

- You might be able to eliminate many, if not all, of the pests simply by removing your plants from the garden and hosing them off in the shower or sink. This might not eliminate all of the pests, but it should wash away many.

If you have a pest problem, then look for these signs that will point to the presence of certain types of pests:

- Webbing in and around the plants – Spider mites
- Silvery streaks on leaves – Thrips
- Black streaks on leaves – Thrips
- Sticky residue on leaves – Aphids
- White clumps and masses on leaves – Mealybugs
- Deformed leaves or stems – Fungus gnats or Aphids
- Yellow or white spots – Aphids, Thrips, Whiteflies, or Spider Mites
- Black spots on leaves – Thrips

Thrips can create an enormous population in a small amount of time, so if their presence is ignored, they can decimate a garden in very little time. These pests are especially attracted to light-colored flowers and plants. Seeing black spots on the leaves is an almost absolute guarantee that you have an infestation of thrips. You might also see where the plants that they have been feeding on will appear dry and have discolored spots.

To get rid of thrips, you can first release some of the insects that will eat thrips like ladybugs or lacewings. If you don't want ladybugs flying around your house, then you might want to think about using a chemical solution.

Mealybugs – These pests love plants that are fruiting, so if you see plants that have yellow, weak leaves, then you probably have an infestation. You will also see cottony-looking white masses on the underneath of the stems and leaves, which are where the eggs are located, and these masses can also be found on other parts of the plant.

These pests can be effectively treated with an insecticide. If you prefer using a more natural method, then you can mix one ounce of Neem oil into one gallon of water and spray the plants weekly until the mealybugs are gone. If you notice an infestation of mealybugs early enough, you might be able to just destroy the egg sacks to prevent them from hatching.

Aphids – These pests will secrete a sticky residue that will stimulate the growth of mold on the plants and will also attract other insects such as ants. Aphids will suck all of the nutrients out of the leaves of the plants and leave them limp and lifeless.

Predator bugs like ladybugs or lacewings will rid your garden of aphids. If you find any type of injury to the plants, those parts will need to be removed even if it is an entire plant. Try to never overfeed your plants as this can attract aphids. There are also soap-based insecticides that will help to rid your plants of aphids.

Whiteflies – These will hide on the underside of the leaves and will fly up in clouds when the leaf is disturbed. They will also leave a sticky substance on your plants and discolored or light-colored spots where the bugs have been eating. You will be able to see the whiteflies on your plants.

Spray your plants with cool water to begin removing the infestation. Predator bugs will also get rid of the whiteflies. The plants can also be sprayed with a solution of Neem oil or organic soap.

Spider Mites – These pests will leave fine lacy webs all over your plants. They particularly like areas with low humidity and high temperatures. They leave behind the sticky webs as regular spiders will, but these webs are finer and smaller. You might also see white or yellow spots on the leaves of the plants because the spider mites will suck all of the nutrients out of the leaves. They will increase their colonies quickly before you see the first web, so regularly check the underneath side of the leaves, since that is where they gather.

To treat spider mites first start by removing all of the infected stems and leaves. Then you can use a plant-safe pesticide to get rid of them. You will also be able to kill the eggs and the mites by spraying the plants with a solution of Neem oil.

Fungus Gnats – These pests are annoying but not actually a huge problem for your plants, other than the fact that they will multiply. The real problem with the fungus gnats is the larvae, since they like to gather near the roots of the plants and feed on them. The earliest sign of an infestation is that the gnats will fly up from the plants in a mass whenever the plant is disturbed. You will find the larvae living in the growing medium.

Don't overwater your plants, as this will attract the fungus gnat. If you find yourself with gnats, try letting the growing medium dry out a bit, as this will kill the larvae since they like moistness. You can also put sticky traps near the growing medium to help catch the larvae, or you can use a Neem oil solution.

PH Balancing

The pH level of whatever medium the plant is growing in will significantly affect the success of gardening and farming. This is true whether the plant is grown in soil or in water, like in a hydroponics system. The pH level of the water is a measure of how acidic that water is. This is important because nutrients are available to the plant at differing levels of acidity. The trick is in finding the exact spot, or very close to it, that the plant feels the most comfortable and is able to extract the most nutrients from the nutrient solution.

To test the pH level of anything, you will compare the reading that you get to a scale with numbers ranging from zero to fourteen. A reading of zero represents the most acidic of the pH levels, and a reading of fourteen represents the most base, or least acidic, of the pH levels. The term 'pH' stands for potential hydrogens, because the hydroxide ions and hydrogen ions are what determine the level of acidity of the water. Testing for the pH level will measure what the concentration of these two substances together is in a solution.

The chemicals hydrogen and oxygen make water. The equation 'H_2O' means that there are two hydrogen atoms for every one oxygen atom in a measure of water. The purest pH level reading is seven, which means that there is an equal number of oxygen, hydrogen, and hydroxide. When you add ingredients to water, you will change the pH of that water, because the added ingredients will breakdown the individual molecules of water, and those will form hydroxide and hydrogen ions. Adding nutrients to the water that you water your plants with will change the pH balance of that water.

Knowing the pH level of your nutrient solution is one of the most important things that you can do as a hydroponic gardener.

The pH number of your water will let you know how soluble (the ability to dissolve) certain nutrients are in the water. The level of the pH will determine how well the plant will be able to suck in these nutrients through the roots and use these nutrients for the health of the plant. Every mineral that you use to feed your plants has its own specific tolerance level to pH. Most macronutrients will not move into the plant if the pH level is too low or too high. This will result in mineral deficiencies in your plants because it will make it difficult for the plants to absorb those macronutrients. The micronutrients that your plant needs will move better at lower or higher extremes of pH. So if the level is too low or too high and the macros are not feeding into the plants, the micros are feeding in at a double time. Either situation will harm the health of your plant. Your plants will either be mineral deficient or mineral toxic.

The ideal pH reading for plants to be able to absorb the bulk of the nutrients that they need is between 5.5 and 6.5. This is also the pH range that is most often recommended for hydroponic gardening. And even though some plants may prefer a slightly different pH level, most plants will be quite safe within this range. Most of the plants that you will grow hydroponically will prefer to have a nutrient solution that is slightly acidic.

You will need to develop a routine for testing the pH level of your nutrient solution, a schedule of times for testing that you will record to see if a pattern of changes emerges. You can choose from several different tools available on the market that you can use for testing the pH level of your solution of nutrients. You can choose from the inexpensive and easy-to-use paper test strips, the liquid test kits, and all the way to the high-tech, relatively expensive, digital meters.

Test Strips

These are the least expensive method for testing the pH level of your nutrient solution. Simply dip the test strip in the nutrient solution and wait for the color on the part that you dipped to change. Then you will compare that color to a chart on the side of the box or bottle that the test strips were sold in. Comparing the color will tell you what the pH level of the water is. This method may not always be the most accurate system, but it is the easiest.

Liquid Test Kits

These kits are also relatively easy to use and just slightly more expensive than the test strips. The liquid test kits will give you a much more accurate reading than the test strips will. To use this system, you will put some of your nutrient solution into a container with a few drops of the liquid test dye drops.

Then you will wait for the color to change and compare it to the chart on the liquid test kit. This method is the most popular among hydroponic gardeners because it is relatively inexpensive but highly accurate.

Electronic Meter

This is the most accurate and most high-tech of the methods of testing the pH of your nutrient solution. The most common of these meters has a pen attached to the box by a cable that is then dipped into the nutrient solution. A reading will appear on the box, telling you instantly the level of the pH in your solution. The digital meters will need to be calibrated occasionally so that they will continue to read correctly, and the pen will need to be handled with care. Even if you opt for the more-expensive meter, it is a good idea to have one of the other methods on hand just in case.

Whichever method you use to test the pH level of your nutrient solution, you will want to set up a schedule for regular testing. It is a good idea to test the solution daily and definitely after adding more nutrients. When the levels of pH in your solution aren't within the ideal range, then it will need to be adjusted to keep your plants healthy. It is a simple matter to lower the pH of the nutrient solution by stirring in some phosphoric acid, or by adding in some potassium hydroxide to raise the pH level.

There are also available solutions that are sold with gardening supplies that will raise or lower the pH level of the nutrient solution. These solutions are simple to use, and they work quickly. They are also much safer to use than the acid method.

The type of hydroponic system that you are using, the climate in the room, the types of plants, the nutrients, and the growing medium that you are using can all change the balance of the pH in your nutrient solution. One of the main jobs you will have as a hydroponic gardener will be to ensure that the levels of pH in your solution are kept within safe levels for the health of your plants. Keeping your levels of pH steady is possibly the most important consideration you will have as a hydroponic gardener.

CHAPTER 6: IDEAL PLANTS FOR THE BEGINNING GARDENER

The hydroponic system can be used to grow a large range of different plants like vegetables, herbs, and fruits. The beginning hydroponic gardener should begin with plants that are easy to grow, since this will help them to become comfortable with the concept of hydroponic gardening. Start with herbs and vegetables that grow quickly, require minimal maintenance, and don't require any special nutrient mixes. You will be able to assess the system that you constructed to see if you actually like the setup, or if you might want to consider another type of hydroponic garden. And choosing plants that grow quickly will allow you to test your system without waiting long months for your harvest. There are three things you will want to pay attention to with your hydroponic garden:

- Remember that the ability of your plants to absorb nutrients needed for growth will require a proper pH level in the nutrient solution.

- The plants will prefer to beat a certain temperature for optimum growth.

- To be able to grow correctly, all plants need the right amount of light.

Hydroponic Herbs

Any gardener would find success with beginning with herbs as their first hydroponic crop. Herbs are easy to grow, and one of the best ways to begin a life of hydroponic gardening. Since hydroponic gardening is built around the use of moisture, herbs that prefer dry conditions will not grow well hydroponically. And there is nothing better than growing the herbs that you can use in your cooking.

DILL – Dill can be used dried or fresh in many different ways. This herb is an easy one to start from seed by simply pressing the seeds into a section of moist rockwool. The seeds will become seedlings in six to ten days and then they can be transplanted directly into the system along with the rockwool. Dill prefers water at a pH level of six to seven. And since this plant will grow to be about three feet tall, you will want to use a grow light that is adjustable, or one that is set vertically next to the plants. When you want to harvest some dill just pull off some of the leafy foliage. Do not pull the stems of the dill off until they are turning brown.

CILANTRO – This herb will grow well from seeds if you press them into any moist medium. Then when they turn into seedlings, you can transplant them directly into your hydroponic system along with the growing medium. Cilantro will be fully grown and ready to harvest in about fifty to fifty-five days and will require very little maintenance. This herb will grow happily in any temperature that is between forty and seventy-five degrees. Cilantro does require fresh air flowing in the room and ten hours of light every day.

ROSEMARY – This is a popular herb in the kitchen, and it is also good for your health. Rosemary will grow more slowly than other herbs will in a hydroponic system. You will have plants that are large enough to harvest in about twelve weeks. Rosemary prefers a pH of somewhere between six and seven and like twelve hours of light each day.

BASIL – This herb can be grown from cuttings off other basil plants or from seeds. If you are starting with cuttings, set them in fresh water until they grow two leaves and some roots, and then set them in the hydroponic system. Basil likes to have at least ten hours of light every day. Basil is another plant that will grow tall, so it will either require adjustable grow lights or lights that are set up vertically. This herb likes room temperatures between seventy and eighty degrees and it also likes fresh air to circulate around it.

If the leaves of the basil plant begin to droop and curl, that is a sign that the temperature in the room is too hot or too cold. Basil also requires a humid environment. The water that evaporates from the hydroponic system should create enough humidity for the basil to grow properly, but if it does not seem like enough, then just set trays of water around the basil to create more humidity.

CHIVES – Chives are the perfect plant to grow in your hydroponic garden for two reasons: they are easy to grow and you can use chives in so many different recipes. You can use chives in any recipe where you would like the flavor of an onion without the strength of an onion, so you can use them in salads, vegetable dishes, soups, and stews. Chives are a small plant and they will grow all year round in a hydroponic garden. These sturdy plants will provide you with a continuous harvest with very little maintenance. You can germinate chives from seeds in less than two weeks by laying the seeds on moist growing medium and keeping them slightly damp. Chives will prefer a growing medium that will hold water around the roots, and keep the air temperature between sixty and seventy degrees. Chives will be mature enough to eat in six to eight weeks if you give them twelve hours of light each day.

CHAMOMILE – This is another herb that is perfect for growing in your hydroponic garden because it is easy to grow and good for you. Chamomile leaves make a wonderful tea that will help you sleep well at night. Lay the chamomile seeds in a moist tray to germinate them. After two weeks, remove and discard the specimens that look weaker and leave the rest to grow into chamomile plants. Place them into your hydroponic garden with the growing medium at the roots. Give the chamomile plant about fifteen hours of light every day and you will have chamomile to harvest in just eight short weeks. When you are ready to harvest the leaves cut about three inches away from the stem and then lay the leaves on a soft cloth to dry in a shady area. Leave some of the flowers left on the stems so that the plant will develop new plants. Store the dried stems and flowers in a dark closet in a sealed glass jar until you need to use them.

OREGANO – This is another great herb for cooking, and you will be able to use your own supplies fresh from your hydroponic garden. Lay the seeds on moist rockwool for two to three weeks to germinate them and then transplant the seedlings into your hydroponic garden. Oregano will grow slowly and it might take eight weeks for your plants to be large enough for harvesting. Give the plants at least twelve hours of light every day but don't let the light come too close to the oregano so you don't burn the plants.

ANISE – When this herb is made into a tea, it will relieve problems with the digestive system like gas and bloating. This herb is also used in cooking to add the flavor of licorice to bread, cookies, and cakes. Anise seeds will need to be germinated in the container they will grow in because the plants are delicate and this makes transplanting them quite difficult. The seeds will germinate into seedlings in two to three weeks. They like the air in the room to circulate but not blow directly on them, and give the plants ten hours or more of light every day. Cut off the top five inches of the stem and hang it upside down when you are ready to harvest this herb. You can keep dried anise in a dark cabinet in a sealed glass jar for up to a year.

Hydroponic Flowers

People, women especially, have been growing flowers hydroponically for centuries. So many times, a woman has left the house of another woman with a cutting from a plant wrapped in wet paper or cloth, a cutting that she then takes home and sets in a glass of water in the kitchen window until it grows roots. This is the simplest and oldest form of hydroponic gardening.

MARIGOLDS – These are some of the easiest flowers to grow in the hydroponic garden because they are so low maintenance.

Marigolds seeds will grow on any type of damp growing medium in temperatures that range from fifty to seventy degrees. And most pests don't care for the smell of marigolds, so they are a good natural pest repellant.

ECHINACEA – This is technically an herb, but the plant does grow lovely flowers. A mature echinacea plant will grow to be between two and four feet tall if you are growing the standard variety. Dwarf echinacea will only grow to be about sixteen inches tall. Echinacea plants need about eight hours of light every day.

CARNATIONS – This is probably the most popular flower to use in arrangements of cut flowers. Carnations come in many different colors and have a soft, sweet scent. Lay the seeds into their growing medium right in the hydroponic system to allow them to germinate. The seeds will germinate in less than three weeks, and they will grow on any type of growing medium. They like five hours of light every day and the carnation will grow to be about two feet tall, so they may need some sort of stake to hold onto.

HYDROPONIC FRUITS – The larger varieties of fruits, especially ones that grow on trees, are not well suited to the indoor hydroponic garden because of the size the mature plant will reach.

But fruits that grow on bushes, like berries, will grow quite well in a hydroponic garden. Keep in mind that fruits grown hydroponically will require more attention than any other crop you choose to grow.

STRAWBERRIES – This fruit is one of the most common choices for hydroponic gardening because it is slightly less demanding than some of the other fruits that you could grow. Strawberries are best grown from the mature runners of other strawberry plants, because strawberries that are grown from seed will not be mature for two to three years. The runners you select should have flowers or at least the buds for flowers. Planting new runners at two to three month intervals will guarantee that you will have a supply of strawberries all year long. Strawberries will grow in any type of hydroponic system, and they particularly like the net pots. Give the fifteen hours of light every day and good air flow.

RASPBERRIES – These berries are a bit tricky to grow indoors but they will yield a reasonable harvest. The larger varieties of raspberry plants are too large for the hydroponic garden, so stick to the small size plants for good growing. Raspberries will need to be planted every year if you want a steady supply of berries, since the plant only lives for two years and produce no fruit during the first year.

If you want your indoor raspberries to bear fruit, then the plants will need to be pollinated by hand. Keep the raspberry plant in temperatures of around seventy-five degrees and give them ten hours of light each day. You will want to buy cuttings from other raspberry plants since they will not grow well from seeds. Raspberry plants will produce a lot of small vines that are called suckers, and these will need to be removed so they don't steal nutrients from the main plant. And all raspberry plants will require some sort of stake or trellis to be tied to.

BLUEBERRIES – This berry prefers to be grown in grow bags because they are susceptible to developing rotten roots if their roots don't have the opportunity to dry out between feedings. Blueberries will need to be grown from cuttings, and the plants will most likely not produce any fruit in the first year of growing. Blueberries require more sulfur than most plants do, and if they do not get enough in their nutrient solution, the leaves will begin to curl inward and will turn yellow. They like no more than sixteen hours of light every day. Blueberry plants will need to have one month of the year where they are kept in conditions that mimic winter, with cooler temperatures and very little light. And blueberries like to have regular pruning, so cut off dead or discolored stems or low growing limbs.

Hydroponic Vegetables

Vegetables are the most popular plants to grow hydroponically for several reasons. They grow relatively quickly, they require minimal maintenance, and they yield a good harvest.

KALE – You can grow kale in any hydroponic system. You can grow kale from seeds or from cuttings off other mature kale plants. Since the cuttings will need to be from only mature plants, it is just as easy to start your kale plants from seeds. Lay the seeds on vermiculite and cover them with more vermiculite. Moisten it slightly and leave it sitting undisturbed until the seeds germinate in about six or seven days. Then let the seedlings grow in the vermiculite for about five more days before you move plant and growing medium together into your hydroponic system. Keep the kale plants one foot from each other so that the leaves will have plenty of room to spread as they grow. The plants will need eight hours of light every day. You will be able to harvest the baby kale in just thirty days, and the more mature variety will be ready in three to four months.

CHINESE CABBAGE (BOK CHOY) – This vegetable matures quite rapidly so this plant is a favorite among hydroponic gardeners.

The first plants will be ready to harvest just thirty days after the seeds are set to germinate, and you will be able to reap three harvests from one set of seedlings. When you are germinating the seedling, they will prefer a half-strength nutrient solution to plain water for moistening the growing medium. Bok choy needs to have air circulating and will need seven hours of light every day.

BEETS – This is a good vegetable to grow in your hydroponic garden, but it will need a bit of special handling. Beets are root plants so they will need to have plenty of growing medium to sit in while they are establishing their roots. You will need to plant them from the very beginning in a container that will be large enough to hold the mature beet plant, since they can easily grow to be two inches in diameter. The container you choose will need to be at least six inches deep and at least four inches square (or in diameter for a round pot). Beets will grow in any type of growing medium you prefer to use. Beets do not like high temperatures and they do well when the air is slightly chilly. Give them eight hours of light each day, starting from the time when you start the plants as seedlings.

POTATOES – This vegetable can be grown from any potato that has sprouted eyes, and the harvest in a hydroponic garden will give you smaller specimens.

Plant the potato as an eye with a chunk of potato attached to it, because a potato will only grow from an eye. Potatoes that are bought in the grocery store have been treated with a chemical that retards the growth of eyes, so it is best to get your potato cuttings from a nursery or another grower. Potato plants will spend most of their time growing covered with growing medium, but they will still require ten hours of light every day. Keep the plants about four inches from each other to allow for growth.

ONIONS – This vegetable is quite easy to grow in any of the hydroponic systems, and onions are easy to grow. Onions from seeds will germinate in less than ten days, or you can give your onions a head start and grow them from bulbs. If you are starting your onions from seeds, then start them on a growing medium that is more of a composite because the seeds are really small in size. Place the seeds in a dark, warm place for the first few days to encourage them to sprout. Onion seedlings do not require any kind of nutrients, preferring fresh water for moistening. And for the first four to five days after you transplant the seedlings into the hydroponic garden, you will not give them any nutrients, only water, because this will encourage the onions to go deeper to seek nutrients, which will give you a larger onion. And onions will need at least ten hours of light each day.

GARLIC – This plant does need a bit of care when grown hydroponically, but it is still relatively easy to grow. The one problem that turns people off from growing garlic is the smell that comes off the bulbs, that smells like, well, garlic! Start growing garlic with just two or three plants, because if you stagger the planting, these will provide enough garlic for the use of most houses without the smell driving you out of the house. It is best to start the garlic plant from cloves of garlic, which is the small inside part of the bulb when you remove the outside layer. Every clove will grow a garlic plant, so separate the bulb into the individual cloves and leave the papery coating on the outside. Plant the cloves in the growing medium with the pointy end facing up, using any type of growing medium, using a pot that is large enough for the garlic to grow in since it is a root plant. Keep the upper tip of the garlic clove just below the surface of the growing medium.

CARROTS – This is one of the easiest root plants to grow in a hydroponic system. Carrots seem to prefer growing in a mix of perlite and vermiculite, although any growing medium can easily be used. Carrots will need to be grown in long thin containers to allow for proper growth. Start the carrots from seeds that you place directly on top of the growing medium in the pot you have chosen. Cover the seeds with more growing medium and keep it moist while the carrots germinate.

This part is a bit tricky because if the growing medium is too wet, the seeds will rot, and if it is too dry, they will not grow. The carrots will send their tops up out of the growing medium, and then you will need to thin out the thin, frail-looking plants. This will happen about six to ten days after planting the seeds. Then you will move the carrot seedlings into your hydroponic system. You will need to keep the top of the growing medium moist so the plant part of the carrot will continue to grow, but the carrot will get most of its nutrients from the nutrient solution below. Carrots want fourteen to fifteen hours of light each day, and they will be ready for harvest in two to three months.

SPINACH – Start your spinach from seeds in the growing medium, where they will sprout in about ten days. Let the seedlings grow for a full two weeks before transplanting them into your hydroponic garden. Thin the seedlings out before you transplant them. The seedlings will not be ready for transplanting until they are about three inches tall and have four leaves. Give the spinach twelve hours of sun every day. Only give the spinach half-strength nutrient solution for the first two weeks after transplanting. Stop feeding your spinach one full week before you plan to harvest it so that it will have a sweeter taste. Whichever plant you decide to grow in your hydroponic garden, you will have the joy of having fresh foods available for you whenever you want them to be.

CHAPTER 7: WAYS TO SUCCEED AND MISTAKES TO AVOID

Hydroponics gardening is an amazing way to grow different types of plants in your home. Hydroponics is rewarding, fun, and challenging. There is more interest in this type of gardening in the last few years than there has ever been before. You can't possibly argue with a type of gardening that will allow you to have fresh produce every day of the year, no matter what the weather is doing outside. With a simple setup on a side table, you will be able to grow enough plants to make a fabulous salad every day. This is also a great way to eat organically without paying the high prices that grocery stores charge for organic produce. This industry is only now beginning to realize the full future potential it has. Hydroponics gardeners enjoy the satisfaction of being able to grow their own food. They will enjoy food that tastes better and will save them money.

Since hydroponic gardening is environmentally friendly, they will be helping themselves and the environment at the same time. Hydroponics is not something to jump into lightly, however. It is a good idea to know all of the things that will help you to succeed and all of the things that you will need to avoid in order to be successful at hydroponic gardening.

Ways to Succeed With Hydroponic Gardening

Use a filter or a screen to put over your exhaust and air intake. With most types of systems, you will want to use some form of submersible pump to push the nutrient solution out of the reservoir and into the growing chamber. You can purchase these pumps at pet supply stores and garden supply stores. If the pump you purchase does not come with its own filter, then you can cut a piece out of an inexpensive furnace filter to use in the pump. You might also want to lay a piece of furnace filter over the drain in the growing tray that leads to the drain tube that takes the nutrient solution back to the reservoir, in order to prevent growing medium from getting into the tube and clogging it up. No one should ever go into your garden area unless they and their clothes are clean. One way to ensure that the area where your hydroponic garden is will stay clean and free from pests is to never go into the room if you and your clothing are not clean. This rule should also apply to guests that want to see your hydroponic garden. Any kind of contaminant that you bring into the garden area, whether on your skin or your clothes, will risk contaminating the plants. In a hydroponic garden, disease and blight will spread rapidly, and could possibly kill the entire garden. So do your plants a favor and never go to visit them if you are not completely clean. Pets do not belong in the hydroponic gardening area.

Dogs and cats especially have no business being near your plants. They carry insect eggs and spores from other plants that your garden plants do not want living on them. Animals might knock over the oscillating fans you are using to circulate the air, or they might accidentally unplug the fans, or the grow lights. And since dogs and cats can't resist eating and drinking anything they think might be appetizing, they run the risk of eating a plant that might be poisonous to them, or drinking the nutrient solution with chemicals that will definitely cause them a bit of indigestion

New plants should be kept in quarantine for two weeks. Many of the infestations that plague home gardens are brought in on the stems and leaves of new plants. All plants being brought into the garden need to be kept in quarantine for two weeks before you put them in the hydroponic garden with the rest of your plants. This will give you time to watch for signs of a pest infestation of any type of disease on the new plants. Your system will need to be cleaned and sterilized between crops. Not only will you need to keep the floor of the hydroponic garden area well swept, and the surrounding area free from dirt, you will also need to clean your hydroponic system in between crops in order to keep the equipment clean and germ free. Don't confuse sanitation with sterilization. Regular sanitation is the vacuuming or mopping to keep the floor clean.Sterilizing is done to the components of the hydroponic system and it is needed for the cleanliness of the plant area.

Sterilization is needed to kill off fungus and bacteria. You will also need to routinely sterilize the tools that you use on your plants. Imagine that you prune a plant that has a fungus with pruning shears, and then you use those same shears on all of your other plants. You have just spread fungus throughout your entire garden. Avoid cross-contamination by sanitation and sterilizing.

Control light and dark with a timer. All plants need a specific amount of light to be able to grow properly. Plants use light in the process they complete to make nutrients for their own use. Plants need light every day, and if you are not using a light timer, then you will need to remember to turn the lights on and off at set times during the day. Plants also require a particular amount of time in which they want to be in darkness. So if you are operating the lights manually, you will need to be on duty every day and every night to be able to turn the lights on and off as needed. If the plants do not get the correct amount of light and dark every day, they will not grow well. Save yourself the aggravation of being tied continuously to your home garden and buy some light timers. Dark time needs to be completely dark. Just like the plants need a certain amount of light, they also need a certain amount of dark.

The plant will not perform the proper processes of making nutrients without regular periods of light and dark. So again, use the light timers.

You will need to change your nutrients and water about every two weeks. The biggest part of being able to grow healthy plants is having clean water, level pH, and appropriate nutrients. While you will probably need to add water to the reservoir in between cleanings, due to the fact that heat and light will cause the water to evaporate, regularly adding fresh water is no substitute for regular cleaning of the reservoir. You will need to keep track of how much water was in the reservoir originally, and how much you have added. When you have added half of the volume of water back into the reservoir, then you will need to change the water, or at least every two weeks. Then you will drain half of the water out of the reservoir and replace it with fresh water. Then you will add in half of the amount of nutrient solution that you used the first time. Then you will stir the nutrients into the water well and test the pH level, adjusting it as needed.

Keep the nutrient solution away from exposure to light. The nutrient solution in the reservoir will need to be kept away from light. Mold and algae require three things to be able to grow: light, food, and water.

The nutrient solution itself provides the food and water, so if you remove the light from the equation, then you will be preventing the growth of algae and mold.

Check the nutrient solution in your reservoir daily for pH levels and adjust it as needed. The nutrient solution pH level is one of the few components of hydroponic gardening that can absolutely make or break you. Keeping a proper pH level will allow the plants to gather the amount of nutrients that they need to grow well and abundantly. Letting the pH level get out of range is a guaranteed way to kill your plants. So test and adjust the level of pH in your nutrient solution daily. Gather all of the equipment that you will need before you begin. There is no crisis worse than being in the middle of anything and suddenly finding out you are missing a critical supply. Or what happens if you suddenly need something in the middle of the night? Most of the tools and components that you will need for your hydroponic garden will be gathered together before you begin to assemble it and put the plants in, but some might be overlooked. You might need to keep an extra bulb or two on hand for your grow lights. Hydroponic systems that use sprayers will require you to keep extra nozzle heads, because these will easily clog with nutrients from the solution and will need to be changed often.

And the vinyl tubing for water flow will sometimes crack and need to be replaced. Keep some extra supplies on hand so your garden will not fail just because you are suddenly missing something important.

Plan and write down the feeding schedule for your plants. All plants will have their own feeding schedule that you will need to follow. Many plants will have the same schedule as similar plants, so they should be planted together to make feeding easier. Even if you are using timers to operate the pump that pumps your nutrient solution to your growing tray, you will still need to keep a log so you know when the plants are supposed to be fed in case something goes wrong. The power might fail, for example, and then you will need to do manual feeding at the regularly scheduled times.

Use commercially available nutrient solutions and follow the directions. Another requirement that plants have pertains to the type and amount of nutrients that they need. All plants have certain food preferences that will need to be followed if you want to reap a good harvest. You can purchase the necessary chemicals and mix your own nutrient solution, which is what many of the large commercial growers do, but for the small home hydroponic garden, it is actually better to buy prepared nutrient solutions.

These commercial solutions will have the right amount of nutrients for your plants along with exact mixing and feeding instructions.

Know what type of light requirements that your plants have and follow their needs. Different plants have different requirements for the amount of light they need daily and the type of light that they like to have. The lights that you use in your house will not be sufficient for the plants to grow. The grow lights that are used for indoor gardens are specially designed to put out certain spectrums of colors that will mimic the light that comes from natural sunlight. Every grow light has its own unique advantages for helping your plants grow. Small fluorescent lights are best for a small space since they don't put out much heat and they can be mounted vertically for tall plants. A high output fluorescent light will grow your plants thick and strong. Fruiting and flowering plants and flowers prefer to use high-pressure sodium bulbs. The light that is most like sunlight is the metal halide bulb. You will need to know what kind of light your plants prefer so that you can set up the correct ones. Learn and understand the type of equipment that you will need for your system. You will need to understand how each of the hydroponic systems works in order to choose the system that best suits your needs.

The space that you are setting up your garden in, the types of plants that you want to grow, and your level of expertise will all determine the size and type of hydroponic garden that you assemble. Don't be afraid to begin on a small scale and learn as you go, and as your plants grow. You can add components to your system later or even build a newer, larger system.

Mistakes to Avoid With Hydroponic Gardening

The second most important part of the hydroponic system is the lighting. If you fail to invest in the correct kind of lighting or don't provide your garden with enough light, then you are setting yourself up for failure. If you buy low-power light solutions or those that are too small, then your plants will suffer. Your plants will not grow if you buy the wrong kinds of bulbs, and the least expensive options might not be the kinds of lights that your plants need. Do some research to learn what lights are the best for the plants that you want to grow.

While your plants need water, it is also possible that you can give your plants too much eater. If you overwater your plants, they will begin to turn yellow, and the leaves will droop, and they might even rot and die. If you notice signs of overwatering just back off the watering for a few days and the plants will eventually correct themselves.

While you are learning about your hydroponic garden, you may not know if your watering schedule is correct. The best way to tell is to stick your finger into the growing medium. If your finger comes out completely dry with no sign of wetness, then it is time to water your garden.

Not all fertilizers are created equally. Conventional fertilizer that is made for adding to soil will not dissolve properly in your reservoir water and the leftover clumps will clog your lines. And these fertilizers do not have the same type of nutrients that are needed in the hydroponic garden. You will need to buy a nutrient solution that is made for hydroponic gardening, and then ensure that it is mixed into the water at the proper level. It is very easy to add too much of the nutrient solution and this will be harmful to your plants. Watch your plants for signs that the nutrient solution is too strong. These signs can include the plant turning brown of wilting. If you see any of these signs, decrease the concentration of the nutrient solution by adding in plain fresh water to the reservoir.

It is critical for your success that you choose the types of plants that you will be able to grow properly. Not every crop will grow exactly the same way in the hydroponic system, and assuming that they do is the fastest way to invite failure. All plants have different needs, and not all plants are suitable for all growing environments.

The fact that you are growing your plants indoors will have something to do with the appropriate plants to choose. You will need to ask yourself if you can grow the plants you chose in your hydroponic garden, and if there are any factors that are beyond your control, that might affect the growth of your plants. You will need to know the needs of the plants that you choose and be honest in examining your system and your own abilities. Don't choose plants that won't help you be successful.

The hydroponic system of your choice can often be assembled with using very little money, but there will be some type of cash outlay in setting up any system. Even if you use all recycled components for the system itself, you will need to, at the very least pay for plants. When new gardeners underestimate the amount of original cash they will need to put out they open themselves up for failure and they may end up with a hydroponic system that they are not able to use.

One of the most critical mistakes that new gardeners make is not being realistic about the space they have available and assembling a system that is difficult or impossible to operate. When growers design systems that do not take into consideration the space available for the garden, then the garden itself is not efficient to maintain.

This can lead to problems where accessing vital components is difficult or impossible, systems that are not conducive to pest control, require too much transplanting of plants or other maintenance, are difficult to harvest the crop, and use the available space in an inefficient manner. When you are designing your hydroponic system you will need to take into consideration the needs you have as the gardener, such as convenience, automation, and easy access to all areas of the system. You will also need to consider the humidity and temperature in the area, the ease of pest control and avoidance, the ability to feed nutrients, the ability to water the plants regularly, and the need for extra lighting.

Common HydroPonic Problems and Their Solutions

SYSTEM LEAKS – Any place that there is a valve or a joint in your system is a place for water to leak out. If roots or growing medium get into the lines and cause a clog, then water will leak out. Always test your empty system with water before you add any plants. Check all of the connections well. Check your system regularly for any type of clogged outlet or drain, or any place where the roots of the plants have become overgrown.

Make sure that the reservoir that you choose will hold all of the nutrient solution that the system will need, not just the amount of solution that will be in the system when it is being used.

LIGHT STRENGTH – If you don't buy enough grow lights, or you buy ones that are not strong enough, then your garden will not provide you with the harvest that you want. Grow lights will usually only provide enough light for a section that is four feet by four feet, so if you have an eight-foot room, then you will need at least two sets of grow lights.

LEARN AS YOU GO – While every plant is different, every crop that you begin will be different from the last one. Sometimes your harvest will go smoothly from start to finish, and sometimes there will be problems that you will need to solve as you go. Always take every opportunity to examine your successes and failures, learn what works well and what needs to be changed, and then make the adjustments that you need to make. Every crop that you grow will be your practice run for the next crop. Use every method of learning at your disposal. Keep journals, logs, charts; take pictures of good plants and bad plants and use them as examples for the future.

MONITOR THE HEALTH OF YOUR PLANTS – You will need to keep a close eye on your plants to keep them healthy. One of the biggest mistakes that novice gardeners make is in thinking that once the plants are planted in the system, and all of the timers are set, then the system can operate on its own. This is definitely not the truth. Watch your plants for signs of distress. Look for signs of pests, deficiency, disease, or poor growth. When a problem is noticed early then, it can usually be corrected, but the problem will need to be noticed.

PROTECT YOUR PH – Not monitoring the pH level daily and adjusting it as needed is another mistake that many new gardeners make. Again, the system can't be left to run on its own; it will need daily attention. The nutrient solution pH level is one of the most important parts of your hydroponic solution. Evaporation, the rate of absorption of the nutrients by the plants, the presence of disease, and excessive temperature can all affect the level of the pH of your nutrient solution. You will need to test the pH at least once every day and adjust it as needed. If you set a schedule to do it at the same time each day, and keep a written log of the results, then it will be harder for you to miss a testing.

USING THE WRONG KIND OF WATER – All water is not the same. If your tap water is hard, then it can cause you to have problems in your hydroponic system. Also, depending on the amount and type of chemicals that are added to your local water, it may not be appropriate for use in your hydroponic system. If there are too many chemicals already present, then you will not be able to add as much nutrient solution as your plants need. And unless you pay to have the water tested yourself, then you might not know the actual concentration of minerals and chemicals in the water supply. The compounds most often present in hard water are magnesium salts and calcium, which are large molecule compounds that your plants will not easily be able to absorb. Any calcium that is present in the water will attract the calcium in the nutrient solution that you add and create molecules that your plants can't use. It is a good idea if you need to use tap water for your hydroponic system, to use some sort of carbon filter to remove chemicals and impurities. It is an even better idea to use distilled water for your system.

NOT HAVING EXTRA COMPONENTS ON HAND – The hydroponics system is dependent on very frequent or constant delivery of nutrients and water to your plants. You can very quickly develop problems if you have a nozzle that is blocked or completely fails, or a pump that stops working.

Systems that depend on a water pump to move the nutrient solution to the growing tray will be virtually crippled if the pump is not functioning. Systems that use spray nozzles are often stopped from functioning if the nozzles get clogged with nutrients. Any time that the water is not able to get to the plants, you run the risk of the roots of the plants completely drying out and dying. If the air pump fails, then there will not be a constant supply of fresh air to the roots of the plants. Vinyl tubes can break or clog. Keep a supply of extra nozzles, water pumps, air pumps, hose clamps, and vinyl tubing on hand.

CHOOSE THE RIGHT GROWING MEDIUM – There are many different growing mediums to choose from, and this is a critical decision for the success of your plants. Some growing mediums should only be used once, and some are reusable after they are cleaned. Some growing mediums will keep the roots of the plants well drained and others will keep the roots of the plants wet at all times. Some growing mediums are inexpensive and some are expensive. Take the time to know what your plants need the most, and then choose the growing medium that will give your plants what they really need.

FILLING AND FLUSHING THE SYSTEM – If you don't change your nutrient solution and flush your system regularly, you significantly increase the chances that you will harm or kill your plants. Water that is not fresh and improperly mixed nutrient solution can lead to problems with pests and diseases. It can be a bit of a job to change the nutrient solution and flush the system regularly, but the effort is well worth it. When you do this, you will find that your plants grow faster and larger, and your harvests will be much more satisfying.

Much of the journey of growing plants hydroponically will involve a good deal of trial and error. Not everything you try will work right, and not every plant you try to grow will grow properly. Many of the issues you could encounter with hydroponic gardening can be completely avoided with careful planning and attention to detail.

CHAPTER 8: THE BUSINESS OF HYDROPONICS

Hydroponic gardening is getting a lot of attention these days and experiencing a boom in interest. These gardens are commonly built-in greenhouses or indoors. Both of these types of farms are proven to be profitable commercially. The hydroponic farm will grow and sell greenery, grasses, fruits, vegetables, and herbs to individuals and businesses. Many smaller restaurants like the idea of buying freshly grown produce for selling to their customers.

If you decide to make a living growing things hydroponically, or even just to make a little money on the side, you will need to take into consideration just what is needed to begin your own hydroponic business. If you intend to set up a large system, then you will first need to choose a space that is big enough to set up a system that will be large enough for the number of plants that you will need. You can rent or buy a greenhouse, or even build a small one on your own land if you have the space and local codes will allow. Actually, setting up the hardware of the business will require some outlay of cash, since once you have the available space, then you will need to set up the systems. You will need components for assembling the systems, racks, lights, fans to circulate the air, and seeds and plants.

You can make money with hydroponics gardening. Start by seeing if there is a local demand for hydroponically grown produce. See what is in demand where you live. By knowing where a need exists, where there is a local shortage, then you will be able to plan to fill that need. If you can fill this need with harvests from your hydroponic garden, you will have found your area. Demand will change constantly during the year, and you will need to be able to adjust your supply to meet that demand. Every plant that you grow will have an offseason, and you must be prepared to change your crops as needed. The small grocer and local restaurant will most likely be your best customers. And don't be afraid to visit these places and ask them what they have trouble getting regular supplies of. If you are able to supply fresh tomatoes or strawberries in the middle of winter, you might be the restaurant owner's new best friend. Sometimes the need is with fresh herbs, like rosemary and basil. You won't know what the local need is until you ask, and then you can plan to grow your crops according to need.

Becoming a local supply of fresh produce means that you will have minimal transportation costs, and your produce will be fresh when it arrives at the merchant. Your regular merchants might even want to come by and visit your operation to see the magic of hydroponics for themselves.

Networking is really important in setting up a local visit, so don't be afraid to ask your regular customers for referrals to other businesses. Visit the locals, the small grocery stores, restaurants, farmer's markets, and local farming shops to know their needs.

The system that you choose to build will need to be as simple as you can possibly make it, both in design and maintenance. Your goal is to be able to harvest as much produce as possible in the least amount of time with the least amount of effort as possible. And by starting with a simply designed system, you will be better able to expand it at a later date if needed. If you are planning to make a home business out of your hydroponic garden, then you will need a system that will let you keep your costs low and keep the system running automatically. You will also want the system to grow healthy produce quickly.

Hydroponics gardening is now being looked at as the best solution to the current need for sustainable gardening. The hydroponic garden is the most likely solution to feeding the ever-growing population of the world. This method of gardening will also allow poorer countries, those third-world countries that are not developed industrially, to be able to feed their residents without needing to rely on sources outside of their boundaries.

This is an important consideration in areas where there is no available land to use for farming, or where the land that is available is rocky or covered with poor soil. One hydroponics farm on a large scale can put out hundreds of pounds of fresh food every year.

People who live in urban areas are also quickly jumping on the hydroponics bandwagon. When you live in an inner-city high-rise apartment building, and you want fresh produce, what do you do? Of course there are local farmer's markets and small grocers, but if you want a fresh supply, you will simply set up a hydroponic garden in an area of your apartment. Many buildings will now allow the residents to set up small operations on the rooftops of their buildings.

By using available technology, hydroponic gardeners can now grow their produce at a much more rapid rate than ever before. Plants can be brought to maturity in a shorter growing period, which will increase the overall productivity of the hydroponic gardens and ensure that even more people will have access to fresh foods. As currently the fastest growing sector of the agricultural word, hydroponics will most likely be the dominating force in the future of food production in the world. Right now, there is only three percent of the land in the entire world that is suitable for growing crops.

Hydroponics gardening will eventually take over a large part of the effort to feed the people of the world. Hydroponics gardening is the new best friend of the agricultural world of the future. You have the ability to get in on the ground floor of a potentially booming business. Even if you only grow enough produce for your own use, you will be contributing to the future of alternative farming.

CONCLUSION

Thank you for making it through to the end of Hydroponics: The Complete Step-By-Step Guide for Beginners to Build a Hydroponic System and Grow Vegetables, Herbs, and Fruits in an Organic Way by Garrick S. Thatcher, let's hope that the book was informative and that it was able to give you all of the tools that you will need to achieve your goals no matter what they are.

Now you will need to decide what type of hydroponic system that you would most prefer to use for your garden and begin to get it assembled. You will need some time to gather together your supplies and assemble your system. You will do some research on different types of plants that you might be interested in growing, and seeing which ones would be the best ones for you to grow.

Use what you learned in this book to get your system set up. You have learned all about the importance of proper pH levels and correctly mixed nutrient solutions. Use all of this information, including the section on the best plants to grow in a beginner hydroponics system, and set up your garden and begin to grow your fresh produce. Pay close attention to the section on the possible components you will need for your hydroponic garden.

Don't forget to keep your system clean and sterile, and to check the pH levels daily and adjust as needed. The information in this book will enable you to be able to start an amazing hydroponic garden.

Finally, if you found this book useful in any way, a review on Amazon is always appreciated!